# EARLY THEMATIC MAPPING IN THE HISTORY OF CARTOGRAPHY

*Arthur H. Robinson*

# EARLY THEMATIC MAPPING IN THE HISTORY OF CARTOGRAPHY

*The University of Chicago Press*

CHICAGO AND LONDON

ARTHUR H. ROBINSON, Lawrence Martin Professor Emeritus of Cartography at The University of Wisconsin—Madison, is the author of numerous works including *The Look of Maps, Elements of Cartography*, and *The Nature of Maps*, which is published by the University of Chicago Press.

THE UNIVERSITY OF CHICAGO PRESS, CHICAGO 60637
THE UNIVERSITY OF CHICAGO PRESS, LTD., LONDON

©1982 by The University of Chicago
All rights reserved. Published 1982
Printed in the United States of America

89 88 87 86 85 84 83 82    5 4 3 2 1

*Library of Congress Cataloging in Publication Data*

Robinson, Arthur Howard, 1915-
  Early thematic mapping in the history of cartography.

  Bibliography: p.
  Includes index.
  1. Cartography—History. I. Title.
GA201.R63        912        81-11516
ISBN 0-226-72285-6    AACR2

*for Mary Elizabeth*

Publication of this volume has been assisted by a grant from the National Endowment for the Humanities.

# CONTENTS

| | Preface | ix |
| | Guide to Notes, References, and Illustrations | xiv |
| 1 | The Thematic Map Appears | 1 |
| | *The Functions of Maps* | 3 |
| | *Revolutions in Cartography* | 12 |
| | *The Thematic Map* | 15 |
| | *The Development of the Base Map* | 17 |
| | *The Technical Milieu* | 22 |
| | *An Overview* | 24 |
| 2 | New World—New Outlook | 26 |
| | *The Scientific Transformation* | 27 |
| | *The Quest for a Universal Standard of Measure* | 30 |
| | *The Growth of Interest in Social Affairs and Statistics* | 32 |
| | *The Organizing of Intellectual Activity* | 35 |
| | *The Growing Interest in the Environment* | 39 |
| | *Industrial and Commercial Development* | 40 |
| | *Epitome* | 43 |
| 3 | From Single Maps to Atlases | 44 |
| | *Edmond Halley as a Thematic Cartographer* | 46 |
| | *Individual Eighteenth-Century Thematic Maps* | 51 |

|   |   |   |
|---|---|---|
|   | *Survey of Early Nineteenth-Century Mapping of Natural Phenomena* | 56 |
|   | *Survey of Early Nineteenth-Century Mapping of Social Phenomena* | 62 |
|   | *Thematic Atlases* | 64 |
| 4 | MAPS OF THE PHYSICAL WORLD | 68 |
|   | *The Atmosphere* | 69 |
|   | *The Oceans* | 80 |
|   | *Geomagnetism* | 83 |
|   | *Geology* | 86 |
|   | *The Land Surface* | 92 |
|   | *Vegetation and Animal Life* | 100 |
| 5 | MAPS OF PEOPLE AND THEIR ACTIVITIES | 109 |
|   | *Population* | 111 |
|   | *Characteristics of Peoples* | 130 |
|   | *Economic Activities* | 140 |
|   | *Movements of Goods and Peoples* | 147 |
| 6 | MAPS OF THE SOCIAL ENVIRONMENT | 154 |
|   | *Moral Statistics* | 156 |
|   | *Medical Maps* | 170 |
|   | *Living Conditions* | 182 |
| 7 | NEW TECHNIQUES AND SYMBOLISM | 189 |
|   | *Techniques of Duplication* | 190 |
|   | *The Production of Tone* | 194 |
|   | *Portrayal of Density and Quantitative Areal Data* | 198 |
|   | *Portrayal of Quantitative Point Data* | 202 |
|   | *Portrayal of Quantitative Linear Data* | 208 |
|   | *Lines of Equal Value: Isolines* | 209 |
|   | *Contours and Layer Tinting* | 210 |
|   | *Curve Lines and Isopleths* | 215 |
|   | *Epilogue* | 218 |
|   | NOTES | 221 |
|   | REFERENCES | 241 |
|   | INDEX | 259 |

# PREFACE

Until recently the study of the history of cartography has tended to be more a history of mapping than of mapmaking. There is a fascination in watching the geography of the world unfold on maps from the earliest times, but a great many scholars (and collectors) lose interest in post-eighteenth-century mapmaking. In addition, there has been a general disinclination to pay much attention to the evolution of the technical aspects of the age-old art and science of mapmaking and to concentrate instead on the geographical content and aesthetic qualities of the maps. These characteristics, along with the natural assumption of "the older the better," have given studies of past mapmaking a strong antiquarian flavor. Witness the following statement by the president of the American Geographical Society in an address entitled "On the Early History of Cartography..." (Daly 1879, pp. 36–37):

> I have not brought the enquiry further down than the time of Mercator (1512–1594), because the progress of cartography since consists mainly in technical details, and is therefore not as important nor as interesting as the previous period.

This paleocartographic bias has had two unfortunate consequences. First, the study of mapmaking since the eighteenth century has been comparatively neglected. This has resulted in overlooking one of the major developments in the history of cartography, thematic mapping, which is concerned with portraying the geographical character of a great variety of physical, social, and economic phenomena. Second, the lack of concern for the methods by which the maps were actually made in any period is a serious omission in a creative field in which the product is a graphic display, the "look" of which is much dependent on technique. This is particularly important in the history of thematic mapping, since a large number of new symbol systems and graphic methods were needed to display the new kinds of data.

Fortunately, in recent decades there has been a growing interest among historians of cartography regarding a variety of significant matters, such as changes in the conceptual approaches to the subject, the development of symbolism, the relation among technical practices and the graphic qualities of maps, as well as the interactions among cartography and individual substantive fields. Unfortunately, however, many scholars in the field have not themselves been mapmakers, and so have tended not to appreciate or even to be aware of the significance of some of these aspects, such as the effect various mapmaking techniques have had on the ability to portray characteristics of data. This was less of a problem in studying the general mapmaking of earlier times, but it assumes major importance in the history of thematic mapping.

Although the roots of thematic cartography can be traced to the last half of the seventeenth century, the critical period of most rapid growth and maturation occurred between approximately 1800 and 1860. As an event in the history of cartography this period ranks with that which occurred in the fifteenth century when Ptolemaic ideas were resurrected. Thematic mapping was a product of the profound developments in all scholarly areas that began in the times of Galileo and Newton, and which by the first part of the nineteenth century had become an intellectual upheaval, marked by the formation of more specialized fields of study, such as geology, biology, meteorology, statistics, economics, and so on. The new investigations required a new kind of mapping which called for new symbolism and new technical methods. Today thematic mapmaking is a major branch of the broad field of cartography.

The subject matter of thematic cartography is almost unlimited. Unlike general maps which simply show where things are, each thematic map is concerned with a particular subject or theme, such as disease or magnetism, many of which are invisible. Their diversity has no limit. To attempt to compile lists of maps in all thematic areas not only would be well-nigh impossible, but would not be very revealing as to the forces and processes involved in the evolution of this class of cartography. Consequently, in order to try to understand its development, it has been necessary to select categories, such as geology and moral statistics, to name only two, that seem representative of both the substantive fields involved and the cartographic symbolism and techniques they required. This necessitated leaving out some relatively large groups of maps that are topically important but not very significant in the evolution of cartography, such as maps of land use, as well as some, such as weather maps, which did not really mature until after the major period of growth had taken place.

*Preface*

Perhaps part of the explanation of why the history of thematic cartography has been rather neglected lies in the fact that the source materials are scattered. Except for a few atlases, most of the early thematic maps appear in a great variety of treatises and scholarly journals. Furthermore, some scholars who have studied the development of particular fields, such as geology, demography, or statistics, have treated the maps as items of secondary interest compared with emerging substantive theory and controversy. Only a few scholars have attempted relatively broad studies; most students have focused on an individual and his contributions. By far the most exhaustive study, and a mine of bibliographic information, is Max Eckert's *Die Kartenwissenschaft*, but even that great work is far from complete for the critical period.

One important related technical matter should be mentioned here. Up to the last quarter of the eighteenth century, maps were printed on rag paper, which not only was durable but tended toward chemical neutrality. With the use of alternate fiber sources and the development of chemical bleaching and pulping, the lasting quality of many papers declined. When they were manufactured these papers were acidic and so contained the ingredient for their own ultimate self-destruction. A large number of the thematic maps made after the beginning of the nineteenth century, especially those in journals and inexpensive atlases, are on paper that is deteriorating rapidly, and some day, unless they are photographed, they will be lost forever.

This book is concerned mostly with the growth of thematic mapping in northwestern Europe, particularly the British Isles, France, Germany, and the Low Countries, and therefore perhaps gives less attention to other areas than they might deserve. This is especially so for the United States. Fortunately, there are a few other studies, such as the extensive historical section of E. Arnberger's *Handbuch der thematische Kartographie*, with its particular attention to Austria, and Castner's "...Search for the Beginning of Thematic Mapping" in Russia. The bias toward northwestern Europe is only partly intentional; some is logistical. There is no question that the first major developments centered in that region, and for such a broad subject the vast resources of the British Museum (later the British Library) Reading Room and Map Room made that great institution in London a natural central base. Extensive use was made of the materials in the Royal Libraries in Copenhagen and Brussels, the Staatsbibliothek in Germany (then at Marburg, now at Berlin), the National Library at Dublin, and especially the Bibliothèque Nationale in Paris. Among the many other libraries and collections which proved useful, particular mention should be made of the Bibliothèque de l'Ecole Nationale des Ponts et Chaussées in Paris and the Ackersdijck Collection of the Library of the University of Utrecht. Needless to say, the reference materials and collections of the University of Wisconsin libraries and the Library of Congress were regularly consulted.

Historical research that involves more than two centuries and is concerned with a great variety of subject matter with widely distributed source materials is heavily dependent upon the contributions of other scholars and the assistance of many people. In a continuing program of research begun more than twenty-five years ago, the number of persons and institutions who have contributed substantially is very large, of course, and adequately acknowledging them all here is simply impossible. Nevertheless, a few deserve mention.

The professional encouragement and unstinting helpfulness of two good friends and outstanding students of the history of cartography, the late R. A. Skelton, superintendent of the Map Room of the British Museum, and his successor Helen Wallis in the British Library have been invaluable. Dedicated scholars, such as Gerhard Engelmann, Cornelis Koeman, John Andrews, and Saul Jarcho have been most helpful, not only by their own published work, but in calling to my attention and making available materials that aided significantly in piecing together the complex story. Equally productive has been my close association with John Fenniman, George McCleary, Karen Pearson, Norman Thrower, and David Woodward, all of whom have carried on intensive research in various aspects of the more recent history of cartography. Karen Pearson read the entire manuscript and made many wise suggestions. Among the many persons in Western Europe who went out of their way to help, a special debt is owed Mlle Monique Carlier of Paris.

When not blessed with a personal fortune, a professor needs assistance in the way of travel and released time in order to carry on a long-term project such as this. The Department of Geography and the Graduate School of the University of Wisconsin at Madison have been generous, especially through the creation of the Lawrence Martin Professorship of Cartography and the research stipend which accompanied it. The John Simon Guggenheim Memorial Foundation granted two fellowships fourteen years apart for the same study project, an act of uncommon trust. I hope that this book, necessarily something of a distillation because of economics, will in some degree keep the faith with these organizations.

Like one who saves the icing on the cake for the last bite, I save the acknowledgment of the more personal kinds of help to the last because, in the long run, it seems to be that upon which I was most dependent. My daughter Patricia Robinson transcribed many research notes, helped with language problems, held pages while I photographed, and was generally a willing aid. For many years June "Sulli" Bennett has been an invaluable "staff assistant," typing, duplicating, and sorting manuscript, handling correspondence and address lists, keeping note-numbering in order and reference form consistent, as well as helping me keep track of a variety of concurrent professional activities.

My wife Mary Elizabeth not only made an immense indirect contribution by being a companion at home and while traveling, but she helped directly by always being available to discuss literary aspects, to read manuscript critically, check final copy, alphabetize various lists, find lost notes, and too many other things to mention. Historical study never really ends, but without her help this research would not be as far along, nor would this book have been possible.

GUIDE TO NOTES, REFERENCES, AND ILLUSTRATIONS

Notes and citations are numbered serially throughout, and all are assembled in one Notes section at the back of the book. Citations of published written works, atlases, and some separate maps are to author and year—for example, Dupin (1827)—and all are arranged alphabetically by author in the References.

The sizes of most maps, where relevant, are given in their captions. The first number is the width as one would normally look at the map, the second is the height. Linear reduction is given as a percentage stating how much smaller than the original is the reproduction. For example: reduced 25 percent = 3/4 of the original; 50 percent = 1/2; 75 percent = 1/4. The method of original production given is the writer's best judgment, but the difference between the results of copperplate and lithographic engraving techniques is sometimes minimal.

Insofar as possible, the years of birth and death of a person are given in the Index immediately following the name.

# *One*

# THE THEMATIC MAP APPEARS

Numerous innovative ideas have advanced the evolution of abstract thinking in man, including, of course, many concerned with geographical comprehension. For example, long before Aristotle assembled arguments to support the notion that the earth was spherical, that fact must have occurred to someone for the first time, and from that moment on, the world of man would never again be the same. A similar achievement occurred when someone first represented geographical facts by arranging signs in two-dimensional space. It makes no difference whether the signs were patterns of pebbles, shells, bones, or marks scratched in earth or clay. The act of mapping was as profound as the invention of a number system. The use of a reduced, substitute space for that of reality, even when both can be seen, is an impressive act in itself; but the really awesome event was the similar representation of distant, out of sight, features. The combination of the reduction of reality and the construction of an analogical space is an attainment in abstract thinking of a very high order indeed, for it enables one to discover structures that would remain unknown if not mapped.

An account of the novel idea of mapping would have been possible only in some ancient saga, since it is likely that maps were made before language marks were devised to represent objects and ideas or the sounds of human speech. We may reasonably assume that long before recorded geography the use of ephemeral maps created by pebble-placing and dirt-scratching was widespread. The art of mapmaking and map using is known to have been practiced among peoples who had had no memory of contact with more advanced cultures and who had not yet progressed to the stage of using a written language. Daly reports that both Admiral Parry and Admiral Ross were amazed to find that the Eskimos not only understood their charts but were also able to continue coastal delineation accurately in areas not then known to the explorers, and Adler's extensive investigation confirms similar talents in other parts of the world.¹ Maps have been termed the oldest of the graphic arts, and as a means of communication about man's living space they have no peer.²

Although there are enough more or less commonplace references to maps in ancient writings to suggest that they were not thought to be anything extraordinary, we actually know very little about the maps themselves. This is primarily because there are almost no extant examples. Only a handful of maps from the Western and Eastern worlds survive from the period before the time of the well-known geographer-cartographer Claudius Ptolemy of Alexandria, about A.D. 87–150. These range from a clay map of northern Mesopotamia of ca. 3800 B.C. and a Babylonian world diagram-map of ca. 500 B.C., to some recently discovered maps in China dating from the second century B.C.³ That there must have been a great number of maps at least of the eastern Mediterranean area and China seems without question because of the relatively good geographical knowledge of the peoples of the area, known from the writings that have survived.

That so few remain is related both to the character of maps as records and to the media on which they were made. There is a natural tendency to discard a map when it is superseded by a better one, and most maps were probably not made to last anyway. Furthermore, interest in collecting maps as records of historical geography did not develop until the Renaissance.⁴ Lloyd Brown points out that no material is known that could be used for mapmaking and nothing else.⁵ Old maps on paper or parchment were good for wrapping things; on stones, for building; on metal, for bullets; or, if the metals were precious, simply for conversion into more convenient shapes and sizes. This vulnerability of map materials to other uses has continued until relatively recent times. Fortunately, the combination of a lesser time span together with a generally steady increase in the total production of maps has resulted in our having today a better representation of the more modern fields of mapmaking in the extensive map collections that have come into being since the Renaissance. The nearer the present the better, and with few exceptions we are not at all handicapped by a dearth of examples from the past several hundred years.

Thematic maps, the subject of this study, have been made only during the past three hundred years. Examples to illustrate their development are readily available, and there is no need to speculate on what they might have been like, as is the case with the maps of antiquity. Furthermore, the mapmakers of recent centuries have often taken pains to write about their cartographic objectives and methods, which with rare exceptions, was not the practice of earlier mapmakers.

## THE FUNCTIONS OF MAPS

Maps are a unique form of communication, and from the very beginning they have been made for some particular purpose or set of purposes. Although they are composed of lines, colors, and other kinds of marks, just as are the wall drawings of cavemen and the paintings of Rembrandt, maps have rarely been made with aesthetics as the primary aim. The representation of geographical space puts many restraints on the cartographer concerning size, shape, and relative prominence, and one cannot shift things about very much to obtain a better balance or a more effective arrangement of colors. The aesthetic aspect has generally been limited to embellishing maps with a variety of decorative features.[6] This is not to say that a map cannot be considered as worthy of wall space as a painting. In many the craftsmanship invites display: the colors may be satisfying, or the area or topic may evoke pleasant memories. But in most cases these attributes are a kind of unplanned dividend beyond the objective that prompted the making of the map.

There are many reasons why maps are made and a host of uses to which they may be put, and to generalize these in but a few categories is to invite disagreement. Nevertheless, to show how thematic mapping fits into the rich history of cartography, a summary is useful. Taking a very broad view, we can recognize that, from early times, maps have served three general functions:

1. As a record of the location and identity of geographical features
2. As a guide for the traveler
3. As a vehicle for the figurative expression of abstract, hypothetical, or religious concepts.

Even though most maps involve combinations of functions in some degree, there are numerous examples of comparatively pure forms. The relative importance of the three functions has varied from time to time, and, as a kind of backdrop against which to view the emergence and maturing of thematic cartography, we shall try to summarize the appearance and roles taken by the functional classes in the age-old practice of mapmaking.[7] It is also important to point out that by its very nature any geographical map, being a representation of geographical space, must locate and identify a variety of geographical features. For some maps that is the primary objective, while for others that serves merely as a background or context.

The references to maps in ancient writings seem primarily to be about the general, or reference, variety (category 1, above). The few maps and fragments that have survived are mostly of this kind and range from showing landownership boundaries (cadastral maps) to city plans (FIG. 1), and even to "world" maps, early ones of which must have been based largely on conjecture.[8] The *Geography* attributed to Claudius Ptolemy, at least parts of which were apparently written in the second century A.D., contains a large body of information concerning the locations of places. No part of Ptolemy's original writings is extant, and it is

**Figure 1** An early Mesopotamian city plan on a clay tablet. (Courtesy of Norman J. W. Thrower.)

likely that the available manuscripts date only from the tenth and eleventh centuries.[9] The maps that accompany these are primarily general reference maps and consist of larger-scale maps of particular areas as well as a "world map" (FIG. 2).

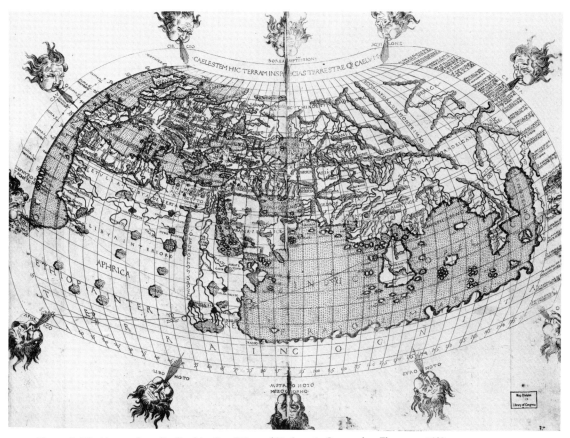

**Figure 2** World map from Berlinghieri's edition of Ptolemy's *Geography*, Florence, 1482. Copperplate engraving. (Courtesy of the Library of Congress.)

Maps made to characterize features and show their locations seem not to have been made very often in the Western world during the Dark Ages, but that kind of objective became more common again after the thirteenth century, and from then on such maps became an increasingly significant part of cartographic production. As the Western world emerged from its parochial isolation and began to explore farther and farther away from Europe, the unfolding world map steadily increased in interest, and the cosmographies or geographical descriptions that superseded Ptolemy's *Geography* were more and more embellished with up-to-date reference maps. By the early seventeenth century elaborate atlases, collections of general reference maps, worked their way into the best-seller category (FIG. 3).

**Figure 3** A map of Portugal by Fernando Secco included in Abraham Ortelius's 1570 atlas *Theatrum Orbis Terrarum*. Original 515 × 340 mm. Copperplate engraving, hand colored. (Author's collection.)

As time went on detailed maps of regions and countries became more and more common and reliable. In the mid-eighteenth century, first in France and then elsewhere in Europe, large-scale plans and topographic maps began to be made. These, the ultimate in the general reference category, fill out this functional class of maps, and today many countries maintain large national mapping agencies that have as primary tasks the preparation and revision of detailed plans and topographic maps.

Maps can serve innumerable practical purposes but, being a reduced representation of real space, an obvious use is as an aid to travel. In ancient times one moved by ship on water, and by cart, camel, or horseback or on foot on land, and we may assume that in those times there were some kinds of sea charts and itinerary or route maps. Almost nothing survived. One that did is known as the *Peutingerische Tafel*, a

Roman–Middle Ages route diagram–map of the region from western Europe to India showing road connections and distances.[10] The map was made on twelve sheets of parchment, presumably for rolling, and when opened out it is 0.34 meter wide by 6.82 meters long. In order to fit that elongated format, north-south dimensions are greatly compressed and the east-west correspondingly extended. Later examples are the road map of Britain (ca. 1350) known as the Gough Map and Etzlaub's ingenious map showing routes to Rome and distances between towns for pilgrims, made about 1500, a jubilee or holy year during which pilgrimages should be made (FIG. 4).[11] Oriented with south, Rome, at the top, the map bears the title in translation, "The way to Rome, indicated from mile to mile with dots, from one city to another through German lands." Route maps have developed along with transportation technology, and today the road map is standard equipment for the automobile traveler.

Although charts to assist travel on water must have been in use, certainly in medieval times and probably much earlier, none dating before about 1300 has survived. Not long thereafter numerous sea charts, now called portolan charts, began to be made.[12] Characteristically they incorporated sets of elaborate wind roses with numerous radiating, intersecting rhumb lines. The portolan charts of the Mediterranean are extraordinarily accurate in their representation of coastal shape and detail (FIG. 5). Charts of the oceans grew steadily better after the initiation of the age of exploration by the Europeans in the fifteenth century and the establishment of national charting agencies, first in Portugal and Spain in the early sixteenth century. As the aids to ocean navigation (such as the chronometer, sextant, and radio) became increasingly sophisticated, so did the maps, and today literally thousands of charts of many varieties are in use by uncounted numbers of mariners, from weekend skippers to the navigators of ocean transports. With the advent of the airplane a new breed of chart, the aeronautical, was added to the travel class of maps.

The first two functional classes of maps, those made primarily to locate and identify geographical features and those made to assist travel, are easy to recognize, and examples are many. The third class, composed of maps made as vehicles for figurative expression, although probably not so large a category as the other two, includes a greater variety. Furthermore, since geography is used as the framework for the graphic display it is not surprising that it is sometimes difficult to distinguish the functions of locating and identifying features from the figurative function of giving substance to abstract ideas and concepts. When a theory of geographical arrangement is itself the concept, the two are completely combined, as was no doubt the case with the early maps of the known earth. Pre-Ptolemaic concepts of geography were

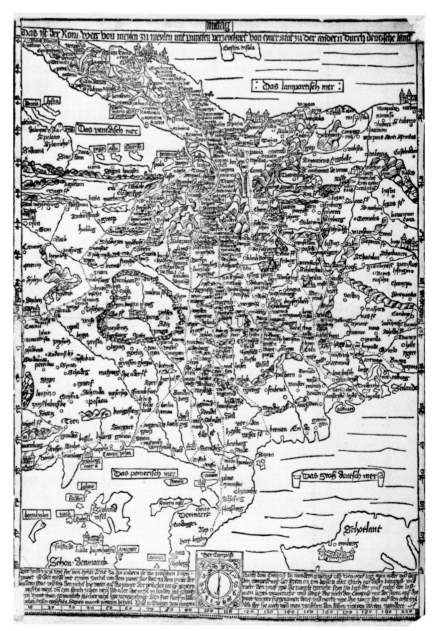

**Figure 4** Erhard Etzlaub's 1500 "Das ist der Rom Weg von meylen zu mylen mit punkten verzeychnet...." South is at the top. Original 289 × 404 mm. Woodcut. (Courtesy of the British Library.)

quite speculative compounds of knowledge with beliefs regarding ideal arrangements and symmetry. Scholarly analyses resulting in hypothetical constructions of how such maps might have looked are the only clues we have, for almost none of the maps exist.[13]

**Figure 5** Mateo Prunes's 1559 chart of the Mediterranean and western Europe. The top of the chart is shown here at the left. Original 660 × 970 mm. Manuscript, color. (Courtesy of the Library of Congress.)

The Christian Middle Ages provide the most abundant examples of this category of maps, largely allegorical. In its early period the church found no fault with the then current cosmology, but as Christianity struggled against paganism it also tended to become antiscience. Since many of the writings of the Middle Ages were ecclesiastically oriented, geographical illustrations of the created earth tended more and more to fit into scripturally related patterns. These were not all alike, but one rather common depiction was in the form of a so-called T–O map which incorporated the concept of a circular earth, inherited from the Roman *orbis terrarum,* divided into three parts as repopulated after the Deluge by the descendants of Noah's three sons, Shem, Ham, and Japheth. Each of the three known continents was assigned to one, and the earth was stylized with east (Paradise) and Asia as the upper half, and with Europe at the lower left and Africa at the lower right. The separation between the upper Asian half and the other two quadrants was along a horizontal line extending from the Don River through the Aegean and continuing along the Red Sea–Nile River corridor. A vertical division between Europe and Africa was provided by the Mediterranean. These two lines formed a T with the junction conveniently and appropriately at the Holy Land in the center of the enclosing O (FIG. 6). A variety of other forms of these *mappae mundi,* some rectangular and some oval, some very large and some very small, appeared during the medieval period, when almost all learning, from cosmography to morality, derived from the church.[14]

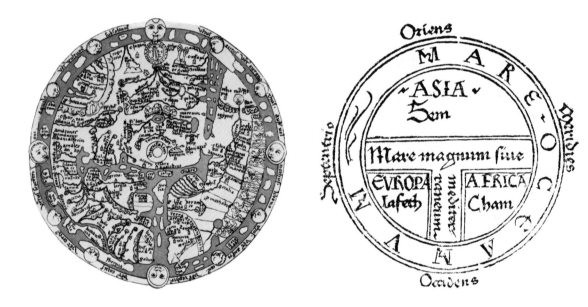

**Figure 6** *Left:* Redrawing of the world map in the Psalter manuscript, ca. 1225, from Konrad Miller, *Mappaemundi, die ältesten Weltkarten. Right:* World map from Isidore of Seville, *Etymologiarum,* Augsburg, 1472, the first printed map in the Western world. Woodcut. (Courtesy of the Newberry Library, Chicago.)

Geographical space was often metaphorically modified, with the Holy Land being greatly enlarged and the lesser-known outer fringes substantially reduced. Much of the symbolism on these maps is a curious mixture of monastic art forms, largely didactic, together with fanciful elements derived from the medieval encyclopedias and bestiaries. The cartographers did not invent the symbols and figures, but simply borrowed directly from the church-oriented symbology of the other visual arts. They employed the same kinds of figurative elements as were used in manuscript illumination and in decoration in church architecture.[15]

Well over a thousand *mappae mundi* were drawn during the period from A.D. 300 to 1500, but almost all of them were incorporated as very small illustrations or as components of the illumination of manuscript writings copied and recopied for monastic libraries. Less than two dozen medieval world maps exist that can stand alone, so to speak.[16] Some of these are quite ornate and studded with all sorts of didactic, allegorical symbols and figures. One such *mappa mundi* is known as the Hereford Map, a large display, 1.3 × 1.6 meters, oriented with east at the top, which was used as an altar piece and is still preserved in Hereford Cathedral in England (FIG. 7). In modern times cartograms and some propaganda maps fall in the figurative class.

Most maps made soon after Europe emerged from the medieval period were based more on observation and reason than on scriptural interpretation, but still they were often prepared to depict concepts at least in part. Many of these concepts are essentially geographical and have to do with the existence of fabled places, such as legendary islands, cities, and water routes.[17] For example, many maps were made to show the location of the hoped-for northwest passage to the Orient from northern Europe. One such is FIGURE 8, which shows unusual cartographic ingenuity in 1578 by using two kinds of lines to distinguish newly discovered areas from those longer known. In describing the voyages of Frobisher, George Beste the author wrote

> I have hereunto adjoyned an universall map,
> wherein my minde was to make known to the eye
> what countries have been discovered of late years,
> and what before of olde time. The olde knowen
> parts have their boundes traced and drawen with
> whole lines, the newe discovered countries have
> theyr bounds drawen wyth points or broken
> lines.[18]

The locations and geographical facts leave much to be desired, but his cartographic heart was in the right place.

By the latter part of the seventeenth century and the early part of the eighteenth, profound changes were beginning to occur in intellectual and technical areas that affected mapmaking in a variety of ways. These will be explored in more detail in chapter 2, but it can be noted here that the consequences in cartography were sufficiently numerous and impressive to constitute what can rightly be called a revolution in the field.

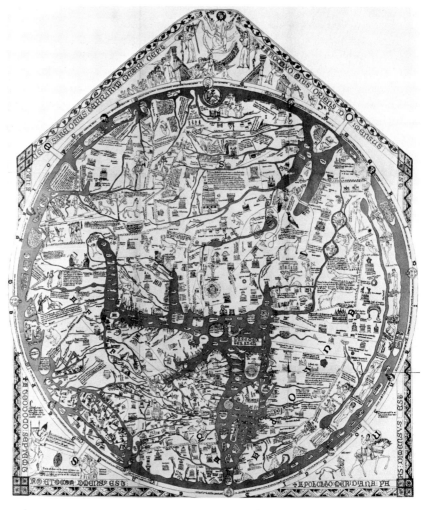

**Figure 7** Redrawing of the manuscript Hereford Map, ca. 1290, from Konrad Miller, *Mappaemundi, die ältesten Weltkarten*. (Courtesy of the Library of Congress.)

## REVOLUTIONS IN CARTOGRAPHY

The changes in cartography since ancient times in the Western world can be visualized as a series of gently sloping surfaces representing periods of little change separated by many abrupt escarpments, each standing for one or more significant developments.[19] Sometimes the surfaces slope upward, as we might represent the period of the increasing knowledge of the later Greek geographer-cartographers, and sometimes the surfaces slope downward, as we would generally characterize the narrowness of mapmaking during the Dark Ages. Most

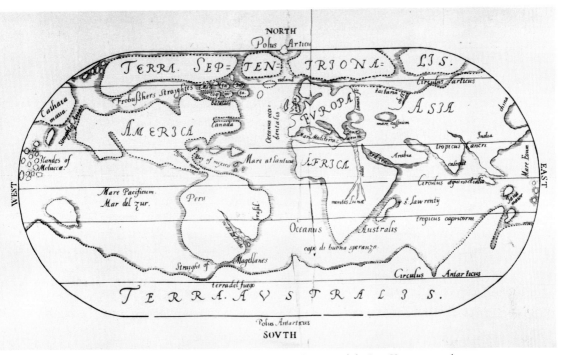

**Figure 8** George Beste's 1578 world map in his *A True Discourse of the Late Voyages... of Martin Frobisher*. Original 385 × 190 mm. Woodcut. (Courtesy of the University of Wisconsin Memorial Library.)

of the scarps in such a characterization of cartographic history are not very high and represent modest innovations, such as the development of a new map projection, the use of stereotypes for lettering on woodcut maps, or the emergence of hachures toward the end of the eighteenth century. Occasionally, however, we see a great cliff representing a relatively short span during which a whole series of major developments took place. These periods clearly qualify as revolutions by fitting precisely the dictionary definition of that term—a sudden, radical or complete change.[20] Staying with the analogy, at the top of such a prominent cliff a new, elevated cartographic plain lay ahead.

The revolutions primarily involve shifts or changes in the intellectual aspects of cartography, namely, the mental models, concepts, or paradigms that motivated the cartographers, and only secondarily the technical operations employed in making the map. The upheavals in the conceptual aspects do not necessarily coincide with the revolutions in the technical aspects, so the precise chronology is likely to appear confusing. One of the major revolutions in the history of cartography is the introduction and development of thematic cartography, the subject of this book, but even though it was largely innovative it was built upon a cartographic heritage. To appreciate the character and significance of the thematic revolution, it is appropriate to look briefly at past major developments in mapmaking.

In the ancient Greek period the development of cartography apparently paralleled the long, steady growth of Hellenistic and Middle Eastern philosophy and science. During the thousand years or so from Homer to Ptolemy, such concepts as the spherical earth, latitude, longitude, map projection, north orientation, and series maps all came into being. These are recorded in the *Geography* of Ptolemy. The maps that must have been made were, no doubt, primarily general maps to locate and identify features. After that, a markedly different period occurred during which mysticism and metaphysics superseded science, largely as a consequence of the dominant role of the conservative church in all intellectual matters. At any rate, mapmaking changed profoundly. With our present overwhelming bias toward science, we usually think of the change as a regression. During that medieval era, to about 1300, mapmaking became very different from what it had been more than a thousand years earlier. So far as we know, the maps were largely vehicles for figurative expression. Suffice it to say that most medieval maps reflect cartographic objectives which are very different from those we today accept as normal. Such was the state of cartography about the thirteenth or fourteenth century when the first of our modern revolutions in cartography began.

It started with one of the great cartographic enigmas, the portolan sea charts, the earliest written record of which is about 1270. Portolan charts were modeled as aids to travel, and they reflect a concern for topological and angular accuracy completely foreign to the other medieval maps. These maps are not on geometric projections, and they did not incorporate latitude and longitude, but their delineation of the positions and shapes in the Mediterranean world is extraordinarily good, especially in view of the fact that there was then no precise way of determining east-west differences. They are indeed a cartographic mystery: no one knows how or when they originated or precisely how their early remarkable accuracy was attained.

The appearance of the portolan chart with its concern for scale and angular accuracy prepared the way for the enormous cartographic upheaval which took place in the fifteenth century. Early in the 1400s refugees moving to the West to escape the Turkish pressure on the remnants of the Byzantine Empire brought along some manuscripts of Ptolemy's *Geography* which had been preserved and possibly augmented in the Arab world. It was translated from the Greek into Latin early in the century, and numerous manuscript copies were made during the 1400s. The *Geography* put mapmaking back on a scientific plane.

It reintroduced a spherical earth, map projections, latitude and longitude, and north orientation, and it even pointed out the utility of separate maps of regions in order to obtain scale advantage. Cartographically Ptolemy was very advanced, actually almost modern.

From a conceptual point of view, in a period of not much more than 150 years, there had occurred a complete transformation. From a view which saw the map as a vehicle for figurative expression with a primary aim of symbolizing spiritually based metaphysical concepts, cartography was turned upside down. The new paradigm conceived of the map as a reduced image of the real earth in the geometric framework of a projection, all done with as much scientific accuracy as observation and knowledge would allow. It is unlikely that there have been many such complete upheavals in any facet of intellectual history. As if that were not enough, the intellectual change was accompanied by a profound technical revolution as well. Soon after the introduction of movable type in the 1450s, the technique of printing was adapted to maps. Within a very few years a large number of maps, including many so-called Ptolemaic maps, appeared in both woodcut and engraved versions.

This first major revolution in modern cartography in the 1400s had deep technical and conceptual consequences that forever changed the field. The conceptual revolution had, in part, shifted the primary objective from the subjective and aesthetic to the scientific, while the technical required trading the freedom of manuscript delineation for the rigid restrictions of mechanical reproduction in order to obtain the advantage of producing numerous copies exactly alike. The fifteenth century must have been an exciting time for someone with imagination and a truly trying time for the reactionary mapmaker.

Although a series of important innovations in cartography which occurred after 1500 produced a number of what might be called minor revolutions, nothing comparable to the profound conceptual and technical changes in the 1400s happened again until near the end of the seventeenth century. At that time the field of mapmaking began to be influenced by two movements which were to have profound and lasting effects. One was the initiation of large-scale topographic mapping and the other was the beginning of what we now call thematic mapping. In a sense, the topographic mapping was merely an extension of the ancient tradition of the map serving as a record of the location and identification of geographical features. On the other hand, thematic mapping was entirely new.

### The Thematic Map

Theoreticians in cartography take delight in trying to define the thematic map, but often get no further than stating that its primary function is not to be a record of the location and identity of geographical features, not to serve as an aid to travel, and not to be a vehicle for figurative expression and allegory. For someone familiar with the entire field of cartography, the negative analysis of eliminating what thematic maps are not serves reasonably well, but it does not help anyone else. Consequently, I shall try to describe this new member of the cartographic family with enough precision so that it is clear why it stands out as different from the other classes of maps.

I pointed out earlier that by its very nature any geographical map must locate and identify geographical features, regardless of its primary function, but that some maps have that as their main function. Such maps are general maps, and they range in type from the small-scale map of a country, a continent, or the world that might have been made by Ptolemy or be found in a modern reference atlas, to the large-scale, very detailed topographic maps or plans produced by the large, national mapping agencies of today. A general map characteristically shows an assemblage of features such as populated places, roads, streams, boundaries, and so on, and displays them in their correct locations so that one may observe their positions relative to one another. Such maps and plans are extraordinarily useful, and when very large-scale and detailed they may serve many purposes.

In contrast to the general map, the thematic map concentrates on showing the geographical occurrence and variation of a single phenomenon, or at most a very few. Instead of having as its primary function the display of the relative locations of a variety of different features, the pure thematic map focuses on the differences from place to place of one class of feature, that class being the subject or "theme" of the map. The number of possible themes is nearly unlimited and ranges over the whole gamut of man's interests in the present and past physical, social, and economic world, from geology to religion, and from population to disease.

An important difference between general and thematic maps and a characteristic of the latter is the portrayal of the variations within a class of features so that the pattern or structure of the distribution becomes apparent.[21] This is considered essential because one of the major reasons for making a thematic map is to discover the geographical structure of the subject, impossible without mapping it, so as to relate the "geography" of one distribution to that of others. This can be done either by correlating the distribution with one's mental maps of other distributions or by literally comparing one map with another. For example, one might portray the distribution of precipitation amounts in order to discover the relationships they might have with the directions of prevailing winds and elevations above sea level, or one might map population numbers to determine how their variations correspond to those of some class of natural resource. Because of the varieties of data that are used for thematic maps, a large number of ways of presenting the data are necessary, and the range of symbolic techniques is very wide.

Before leaving this short review of the nature of the thematic map, it is important to point out that many maps combine functions and that therefore there are many maps that are partly thematic and partly general. There seems to be a natural tendency among mapmakers to try to make their maps as useful as possible, but unfortunately the objectives of portraying clearly the geographical structure of a distribution while also providing a goodly amount of reference data seem to be essentially antagonistic. A good example is provided by a geological map; the more detail regarding the lithologic character of the rocks, the less apparent are the basic structural relationships.

In a sense, general maps and thematic maps are at the ends of a continuum, at one end of which is the objective of displaying simultaneously a variety of data, and at the other end of which is the objective of portraying the structural character of a single class of data. Most maps turn out to be a compromise. A good example is provided by the display of landform data which has always been important in mapping. An array of numbers showing elevations (spot heights) can provide considerable information but hardly any expression of the three-dimensional, structural relationships composed of the slopes, hills, valleys, ridges, and so on. The more effectively these are portrayed by contours, hachures, or plastic shading, the more thematic is the display, regardless of scale or the basic class of map on which it appears.

No map which is primarily thematic appears to have been made before the last half of the seventeenth century. To be sure, occasional "thematic" additions had been entered on otherwise general maps, but the idea of making a map solely for the purpose of showing the geographical structure of one phenomenon seems not to have occurred to anyone.[22] It was not until some exceptional individuals became inquisitive about the earth, and later about its inhabitants and their activities, that their curiosity resulted in thematic maps. How this came about and then evolved, who were the main characters in its development in Western Europe, and what were the relationships among them will be treated in subsequent chapters. Suffice it here to emphasize that the development of thematic mapping in the Western world ranks as a major revolution in the history of mapmaking. Its intellectual and conceptual consequences are comparable to those that followed upon the spread of the concepts in Ptolemy's *Geography* some three centuries earlier.

THE DEVELOPMENT OF THE BASE MAP

Most thematic mapping requires accurate base mapping. Precise, large-scale mapping and charting are prerequisite for two reasons. In the first place, the mapped information, whether it be population, geology, or ocean currents, must be displayed on a base map. Second, much thematic mapping uses statistical data which is related to internal boundaries for its enumeration as well as its portrayal. Accurate and detailed maps are required for both aspects. During the period from the mid-seventeenth century to the mid-nineteenth, the evolution of the thematic map was clearly a fundamental new development, but it took place while the field of general mapping was also experiencing important changes.

The developments, although simultaneous, had little connection with one another. With only a few conspicuous exceptions, such as Sir Thomas Larcom in Britain, thematic cartographers had no official connection, and little professional contact, with the makers of general maps. Obviously each group knew, at least in a general way, what the other was doing, but although the thematic cartographer derived great direct, practical benefit from the activities of the topographic mapmaker and hydrographic chartmaker, the general mapmakers had only an intellectual interest in thematic mapmaking.

In the mid-seventeenth century general maps and charts were poor. The maps of land areas of any extent were usually compilations from a variety of maps of smaller areas of doubtful validity, while most charts of the oceans and their coasts were even less correct. Accurate reckoning of position, distance, and direction were next to impossible, and this constituted a most important deficiency, since it made sea travel exceedingly hazardous. Especially important was the fact that there was no reliable way of determining longitude. Governmental powers began to appreciate the extreme value of accurate maps and charts, and it was France that took the lead in doing something about it.

The minister for home affairs in the France of Louis XIV was Jean Baptiste Colbert, an amateur scientist who realized the importance of the potential contribution of scientists to the state. Colbert was able to arrange the establishment in 1666 of the Académie Royale des Sciences, to which he tried to draw scientific leaders of Europe by offering generous stipends and the advantage of working in Paris at the most brilliant court in Europe. Although scientists in all subject areas were involved, one of the major aims in founding the Académie was to correct and improve maps, charts, and navigation, and this called for improved astronomic and geodetic activity, ranging from studies of the pendulum and the telescope to the actual measurement of the earth. The most important task was to find a way to determine longitude, and the solution turned out to be a combination of the perfection of the pendulum clock and the observation of the satellites of Jupiter.[23] The former resulted from the ingenuity of the Dutch mathematician, astronomer, and physicist Christiaan Huygens, and the latter from the observations of the Italian astronomer and geometer Giovanni Domenico Cassini. Both Huygens and Cassini were lured to the Académie. Huygens returned to the Netherlands in 1681, but Cassini stayed and created a veritable dynasty of geometers which guided French large-scale mapping for more than a century. This consisted of himself, known in France as Jean Dominique, his son Jacques, his grandson César François, also known as Cassini de Thury, and Jacques-Dominique, his great grandson. The Cassinis led the way in the innovations in general mapping.

In bare outline, the survey-mapping activities during the last half of the seventeenth century and into the first half of the eighteenth included the accurate survey of France, the construction of detailed charts of its coasts and harbors, and the determination of the size and shape of the earth. It all began with the extension of the survey of the meridian of Paris by Picard and the triangulation and the preparation of a detailed map of the Paris region which was finished in 1678 at a scale of 1:86,400, one ligne representing 100 toise.[24] Detailed charts of the coasts and harbors were completed and published in 1693 as an atlas, *Le Neptune François,* which provided France with official charts based on official surveys.[25] The further extension of the meridian of Paris was interrupted by budgetary problems and war, but it was eventually completed along with other triangulation in France, resulting in a considerable revision of the outline of France (FIG. 9).

One of the interesting consequences of the measurement of the lengths of degrees of latitude along the Paris meridian, from which the size and shape of the earth were calculated, was that the earth appeared to be flattened a bit near the equator and to bulge near the poles. This was contrary to the gravitational theory of Isaac Newton, and to settle the matter once and for all French expeditions were sent to Peru and Lapland to survey with great care the lengths of arcs of the meridian. The polar degree turned out to be the longer, and Newton was confirmed. Another obvious consequence of that finding was that the original survey of the Paris meridian and much of the rest of France must therefore not be accurate and so had to be redone.

In 1747 work began on a new survey and map of all of France, also at the scale of 1:86,400, under the direction of Cassini de Thury.[26] It was completed for all areas except Brittany by 1789 before the disruptions of the Revolution, but Cassini de Thury died in 1784 and the work was carried on by his son Jacques-Dominique. As a topographic map it left a great deal to be desired in the way of representing the landform, but it finally put things in their right places. Its official name is "Carte géométrique de la France," but is more appropriately known as the "Carte de Cassini" (FIG. 10). It was a monumental achievement, the value of which was quickly appreciated by other countries, and official surveys were soon established by most nations. By the early part of the nineteenth century the general map of Europe had taken shape, and a proper base was finally available for gathering data and for the thematic maps then beginning to come. On the other hand, with few exceptions the rest of the land areas of the earth remained far behind until well into the twentieth century.

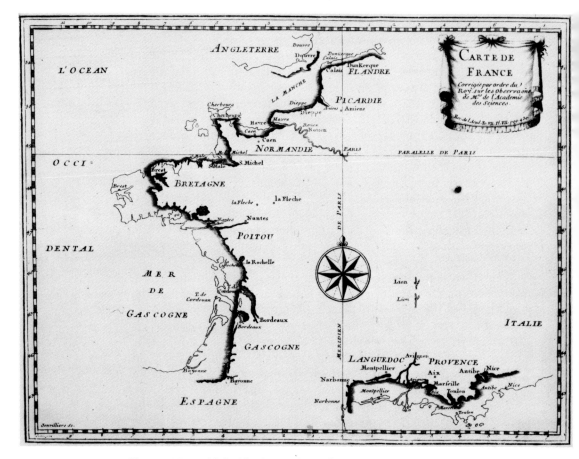

**Figure 9** Map published by the Académie showing the 1693 outline of France *(shaded)* as a result of survey compared with its delineation on a 1679 map by Sanson. (Courtesy of the University of Wisconsin Memorial Library.)

Mapping the ocean base posed a different problem from mapping the land, because, though longitude on land could be ascertained with a pendulum clock and observation of Jupiter's satellites, the use of a pendulum clock on a ship was out of the question. The coasts could be put in the right places by surveys on the land, but making observations out in the open ocean of currents, winds, compass declination, and other such phenomena of interest to mariners and thematic mapmakers

# The Thematic Map Appears

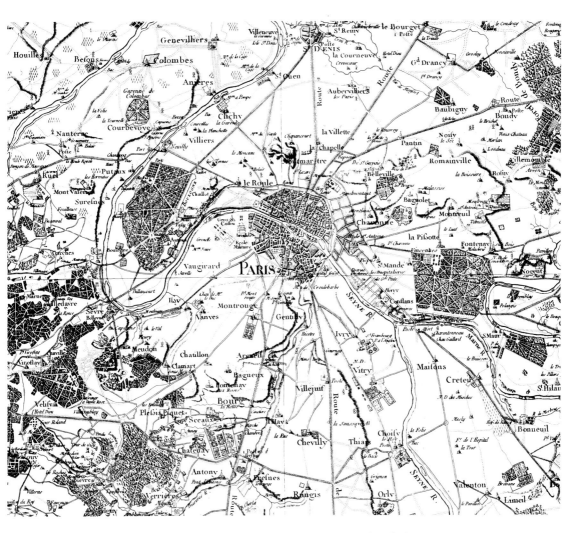

**Figure 10** A section of the "Carte de Cassini." Original 1:86,400, reduced 50 percent. Copperplate engraving. (Courtesy of the University of Wisconsin–Madison Map Library.)

was difficult, to say the least. North-south latitude could be ascertained by celestial observation—not a complicated operation and quite reliable. On the other hand, east-west longitude could be estimated only by dead reckoning, that is, calculations based on judgments of speeds and directions, not a very sure procedure in view of the unknown currents and compass declinations. The perfection of the chronometer, an accurate spring-powered clock, by Harrison and others in England

and Le Roy and others in France, in the latter half of the eighteenth century, finally made the determination of longitude quite reliable.[27] This allowed the mapping of some ocean-based thematic phenomena to become trustworthy by the early nineteenth century.

Although a reasonably accurate base map became available for large areas, the gathering of thematic data was inconsistent for both the oceans and the land. Data for well-known land areas could be depended upon, and so could data obtained for the much-traveled shipping lanes on the oceans, but very large areas of the oceans were essentially unvisited. Since nearly three-quarters of the earth is covered by the seas, thematic world maps of many geophysical phenomena were cartographic estimates at best.

### The Technical Milieu

Making any map calls not only for obtaining data but also for designing and executing the graphic means by which the data are portrayed. In the period from the mid-seventeenth to the mid-nineteenth centuries when the thematic map came into the cartographic picture and grew to maturity, almost all maps were made to be reproduced. The duplication processes placed significant constraints on the graphic character of the portrayal, both because of the physical limitations of the techniques and because of the costs involved. While thematic mapping was developing, changes were also occurring in the methods available for duplication. A brief survey is in order to provide a background for the production problems facing the early thematic cartographers.[28]

From the time of the first printed maps in the late 1400s until the early nineteenth century, there were two processes available for duplicating maps. The most common at first, woodcut, was a relief process in which the image was left standing on a wood block by carving away all the nonprinting parts. Just as in letterpress, the raised surface was printed by inking it and pressing it on paper. The great advantage that a woodcut could be printed along with the letterpress for a book was increasingly outweighed by the facts that the image was coarse, tones were difficult to obtain, and lettering was a serious problem.[29] A much later variant of the woodcut process called wood engraving, in which the relief image was produced on the end grain of a wood block with engraving tools, was used for many simple, small maps that accompanied letterpress in schoolbooks, encyclopedias, magazines, and newspapers in the nineteenth century. They did not play much of a role in the development of cartographic technique, but they did help expand the clientele for thematic maps. In the period from 1650 to the mid-nineteenth century very few woodcut and wood-engraved thematic maps were made, for they were greatly outnumbered by maps made by two other processes. One of these is an intaglio process, in which the printing image is incised in a polished metal plate, almost always copper.

Copperplate engraving had many advantages over woodcut: the lines were more precise and delicate, tones were easier to obtain, lettering was not difficult, and copper plates lasted longer than wood blocks, could be larger, and could be altered with relative ease.[30] Maps printed from copper plates did have drawbacks. If they were to accompany a written study they had to be printed separately from the letterpress and then tipped into the volume, which was expensive. Furthermore, they were inherently costly to produce.[31] The expense probably tended to restrict the number of thematic maps that were made, especially in the early years of thematic mapmaking when there was little ready market for most such maps compared with that for general maps.

Lithography, a new printing process based upon the fact that oil and water would adhere to a particular smooth limestone but not to one another, was discovered at the very end of the eighteenth century.[32] An image receptive to oily ink would be created on a slab of limestone, and if the stone were then successively dampened, inked, and pressed against paper, only the ink image would be transferred. The operations could be repeated over and over again to obtain additional copies. Devised in Munich and first called *chemische Druckerei* (chemical printing) by its originator, Aloys Senefelder, it soon came to be known as *lithographie* (stone writing) in France. Even though metal plates are now used instead of stone slabs, the process is still called lithography. It is a planar process, since the image is on an essentially flat surface. Lithography had several advantages over the relief and intaglio processes, especially that it was quicker, easier, and cheaper.[33] The development of the process in terms of versatility and quality was steady, and a large share of the thematic maps of the first half of the nineteenth century were lithographs. By far the most important factor in its favor was expense, since quite adequate quality could be obtained by lithography at a fraction of the costs of copperplate engraving.

Much of thematic cartography demanded tones, shading, and colors to portray the qualitative and quantitative differences with which it was concerned, such as different geological formations or vegetative regions and geographical variations in the density or frequency of various phenomena such as population or disease. Mezzotint rockers, mechanical ruling, and aquatint etching were used to produce tones and shading in copperplate printing, while crayon shading and mechanical ruling could accomplish the same effects in the faster lithographic technique. Improvements were rapid during the first half of the nineteenth century, and by about 1860 a cartographer was not greatly constrained in this important aspect of cartographic design.

The application of color to maps by printing processes was uncommon until after 1850 for two reasons. In the first place it was expensive, since each additional printed color required its own plate and a separate trip through the printing process. In the second place the accurate positioning of successive colors, called registry, was difficult both because precisely matching images were hard to draw and because the dampened paper employed in both copperplate engraving and lithography varied in dimensions with printing pressure and moisture content. As a consequence, the age-old practice of coloring maps by hand persisted, since it was cheaper and the results were more sure. As the complexity of thematic maps increased, the desirability of using color also increased. Also, the number of maps needing color rose to the point where it was too slow and expensive to accomplish the coloring by hand. Advances in printing techniques made printed color increasingly common after the 1840s and not at all unusual by 1860.

As one looks at the two-hundred-year period during which thematic cartography grew and came to flourish, one is impressed by the slow technical development of the production processes during the first three-quarters of the period and the rapid changes that took place in the first half of the nineteenth century. A printing historian sums it up as follows:

> The man with a mind to improve printing in 1810 thought in terms of wooden presses barely changed for three hundred years, hand-cut blocks manually inked from leather covered pads, and a maximum rate of about two hundred impressions an hour. Fifty years later the inventor had at his disposal photography, electrotyping, fast steam presses with automatic inking and printing rates of some thousands an hour.[34]

### An Overview

As we shall see in subsequent chapters, the increased technical tempo in the first part of the nineteenth century was matched by innovations in thematic concepts and symbolism, making the first sixty years or so of the nineteenth century the period of most rapid development of thematic mapping, unparalleled for at least another hundred years. That is not to say that technical advances did not continue in the later years of the nineteenth century, but primarily they made things easier, faster, and cheaper rather than initiating any fundamental changes in thematic cartography.[35]

Although the first five or six decades of the nineteenth century were a kind of "golden age" of thematic cartography and are the steepest part of the escarpment representing the thematic revolution, its roots go back to the late seventeenth century. The explosion of new ideas and methods that came after 1800 should not push into the background the earlier developments in thematic mapping that were profound in the intellectual and cartographic milieu of their period. An orchestra swelling to a crescendo is impressive, but so is the sound of a single flute when everything else is quiet.

The thematic map is a reflection of man's curiosity about the nature of a great variety of geographical phenomena and the interrelationships among them. That curiosity is a function sometimes of practical need, sometimes of intellectual inquisitiveness, but to satisfy that curiosity the subject must be portrayed on a map so that one not only can see the structure of the distribution but can scan the display in search of correlations and explanations. The practical needs and the inquiring minds which nurtured thematic mapping evolved over some two centuries, and in the next chapter we shall survey the exceedingly complex set of relevant conditions and events from the mid-seventeenth to the mid-nineteenth century.

# Two

# NEW WORLD—
# NEW OUTLOOK

If one could survey all the thematic maps ever made arranged in chronological order, he would be impressed by three aspects of the array. Most obvious would be that very few were made before 1700. Second, the variety and sophistication of the thematic maps of today would appear remarkable; the present capability for symbolism and production is more than adequate to portray almost any combination of data form and purpose. The third aspect, which would be less apparent except on closer scrutiny, is that the objectives of thematic cartography and the graphic means to meet them had nearly all come into being by about the mid-nineteenth century. In Western Europe the critical period during which this new kind of cartography was born and matured lies between the mid-seventeenth and the mid-nineteenth centuries.

The reasons why thematic mapping began and progressed so relatively quickly and in such an impressive fashion are complicated. General maps have been around in one form or another for several millennia at least. The urge to record geography has always been with us, but it had been limited largely to the relative positions of individual things such as coasts, islands, and cities. No concern had been felt to record and display the other aspects of the earthly environment, such as the array of its rock forms or the variations in the mysterious magnetic forces, to say nothing of complex and abstract ideas regarding moral qualities of populations. Such kinds of maps demanded a different level of conceptualization, which could come only with a new way of thinking and a change in attitude toward the earthly home of man.

In western Europe there had been numerous and quite sophisticated technological developments since the time of the relatively primitive Greeks and Romans. They had utilized little rotary motion, did not harness draft animals in front of one another, and their buildings, regardless of their "classic" design, were technically simple compared with the complex engineering required for medieval cathedrals.[36] In contrast to the great technological strides, no significant similar development had taken place in intellectual matters. By the beginning of the seventeenth century, ideas about nature were still essentially Aristotelian, dominated by ideas of perfection, nonmathematical description, and a lack of experimentation. This began to change in the seventeenth century, and out of this change grew the attitudes and interests which called for a new kind of mapping. Thematic mapping was to be concerned with a wide range of subject matter, so it is to be expected that developments in many fields would be involved in its growth.

We shall survey the transformations in intellectual, social, and economic affairs and attitudes which occurred during these revolutionary centuries that were of particular significance to mapping in order to make clear how dependent on them the rise of thematic cartography was.

### THE SCIENTIFIC TRANSFORMATION

The fifteenth and sixteenth centuries have been called the Age of Exploration. Columbus bumped into the Americas; Diaz, da Gama, Cabral, and others sailed around Africa; and by 1522 one of Magellan's ships with a handful of survivors finally circumnavigated the sphere. By the mid-sixteenth century most of the lands previously unknown to Europeans had been visited, and numerous expeditions had been mounted to explore the interiors of the new areas.

Exploration was not confined to the earth. Early in the seventeenth century Galileo, using his newly devised telescopes, discovered the satellites of Jupiter and the rings of Saturn and established that the surface of our moon was not smooth. By demonstrating the validity of the Copernican heliocentric system, he released the earth from its Ptolemaic position at the center of the universe, and in a real sense he set in motion an endless chain of enlightening developments that is still going on.

Isaac Newton was born in 1642, the year Galileo died, and by the end of the seventeenth century the intellectual world had experienced a series of discoveries in astronomy, mathematics, and abstract thought as exciting as the geographical discoveries of the previous century. The publication of Newton's *Principia* (Mathematical Principles of Natural Philosophy) in 1687 was, in a sense, the culmination of a series of developments in dynamics, mathematics, and natural philosophy, the name for the study of nature in general. The earlier insight of Copernicus, Kepler's discovery of the planetary motions, Descartes's analytic geometry and philosophy, and the contributions—and disagreements—of many others, such as Pascal, Huygens, and Torricelli, all came together in a new way of looking at nature.

The law of universal gravitation was a revolutionary, quantitative description of observable facts related to basic laws of motion. At the beginning of the seventeenth century the universe fit the earth-centered medieval concept of immutable circular motions and celestial perfection with the planets influencing "men's temperaments and terrestrial fortunes while at the same time declaring the glory of God."[37] At the end of the century the rigid, perfect universe—including the earth—had become available for observation, and the search for its character and the influences which mold it had begun. The ascendancy of observation and reason over the a priori assumptions of scholastic thought came relatively quickly in England, but somewhat more slowly in France. There the new attitudes, known only to a few advanced thinkers in the early eighteenth century, were popularized by Voltaire in his *Lettres philosophiques* (1734) and more specifically by his *Eléments de la philosophie de Newton* (1738).

One who surveys the intellectual developments from the latter part of the seventeenth century on is impressed by the increasing tempo of marked advances. Probably the most notable event was mathematical, namely the invention of the method of "fluxions" by Newton and "infinitesimals" by Leibniz, later systematized in what we now call the differential calculus. This permitted the study of the rate of change of something as influenced by the variables on which it is dependent. This enabled, for example, Clairaut in his *Théorie de la figure de la terre* (1743) to investigate the form a fluid would assume rotating around its center of gravity, to establish finally the spheroidal shape of the earth, and to substantiate Newton's mathematical reasoning. A second area of mathematics which was to have increasing significance also began in the latter half of the seventeenth century, namely the study of probability. From this was to stem the development of statistics as a field of study with which the new cartography was strongly associated. By the early nineteenth century it had become well systematized, culminating in Laplace's *Théorie analytique des probabilités* (1812).

During the eighteenth century chemistry also took on a new image. The recognition of several kinds of gases such as oxygen, hydrogen, and nitrogen by Priestley, the discovery of many new elements, and Lavoisier's new theory of combustion, replacing the phlogiston theory, transformed alchemy into a precise science in which mathematics played a significant role. The recognition by Haüy of the mathematical basis of crystal forms and the studies of Berzelius into the chemical characteristics of minerals brought system into this important part of the field of geology.[38] Geology provides a good example of the remarkable developments that took place in science during the two centuries that concern us.

From the mid-seventeenth century when Steno advanced the idea of horizontal deposition and superposition, geology—or geognosy as it was early termed—has been of great interest to both professional and amateur scientists. Because it is concerned with the earth and varieties of distributions on it, geology was one of the earlier subjects in the development of thematic cartography. It seems to have had more than its share of controversy, probably because the study of the rock formations leads one to speculate regarding their origins, and then inevitably even regarding the beginnings of the earth itself. The problems involved in reconciling what was observed in the field with what was interpreted from the Scriptures led to some bizarre theories. One such, which may have had its origin in the earlier idea of the depositions of stratified rocks during the Deluge, was that of the remarkable teacher Abraham Gottlob Werner of the Mining School at Freiburg in Saxony. Werner asserted that all rocks were deposited in four successive series—Primitive, Transition, Stratified, and Alluvial—out of a primeval ocean which completely covered the earth, and that the composition of those rock types was related to their age.[39] Field observation by William Smith in England and Georges Cuvier in France in the early nineteenth century demonstrated that fossil assemblages provided a far better key to identification. This promoted a whole new outlook when combined with the ideas of Hutton, a Scot, who asserted that present processes were sufficient to account for the nature of the surface character of the earth, but which demanded far more time than the six thousand years allowed by literal biblical reckoning. By the end of the first half of the nineteenth century, geology was ready to proceed in modern scientific fashion, at least in terms of chronology and identification.

The development of geology and mineralogy from the seventeenth to the nineteenth century exemplifies another change in the interest in natural phenomena that took place during that period. Beginning in the seventeenth century, the general attitude toward nature was rather abstract and disconnected, in the sense that there was a concern with the mathematical analysis of the planetary motions, with the shape and size of the earth, with the nature of minerals as examined in the laboratory, and so on, as when the various aspects of nature are observed and analyzed as separate phenomena. As the period wore on, however, a new attitude developed in which nature was examined in place, so to speak, and concern turned to the forms and combinations of the things that occur in the environment. The earlier, more abstract view, has been termed that of natural philosophy and the later has been called a morphological view of nature, or natural history.[40] The earlier contributions of Werner, Cuvier, Hutton, Smith, and others reached a climax in the work of the German geographer Alexander von Humboldt, a "morphologist of nature" on a grand scale, whose studies of geophysical and biological distributions were carried out over large areas.

These have been called "Humboldtian sciences" and described as follows:

> the great new thing in professional science in the first half of the 19th century was Humboldtian science, the accurate, measured study of widespread but interconnected real phenomena in order to find a definite law and a dynamical cause. Compared to this, the study of nature in the laboratory or the perfection of differential equations was old-fashioned, was simple science concerned with easy variables. Insofar as you find scientists studying geographical distribution, terrestrial magnetism, meteorology, hydrology, ocean currents, the structures of mountain chains and the orientation of strata, solar radiation; insofar as they are playing around with charts, maps, and graphs, hygrometers, dip needles, barometers, maximum and minimum thermometers; . . . They are eagerly participating in the latest wave of international scientific activity.[41]

The growing interest in natural history had, as might be expected, a profound influence on the development of thematic mapping. Just as the examination of tiny phenomena requires enlargement by a microscope to make them visible, the larger distributions in nature need to be reduced by the techniques of cartography to bring them into view. Those who studied such things as the distributions and interrelationships of rock types, terrain, climatic factors, vegetation, and so on, soon found that maps were essential for their study and for the portrayal of their findings.

Before concluding this cursory look at the enormous developments that took place in science during the two centuries, it is well to remember that there were many incongruous associations among beliefs and observation. For example, Sir William Herschel, a natural historian of the heavens, through prodigious use of the telescope (he made his own) by the early nineteenth century had cataloged 2,500 nebulae and hundreds of stars and had discovered infrared rays. Yet he held the view that the sun was inhabited, its population being protected by a cloudy veil from a hot outer shell.[42] Such attitudes were not limited to individuals; near the end of the eighteenth century the French Académie vehemently asserted that the fall of meteorites was impossible.

### THE QUEST FOR A UNIVERSAL STANDARD OF MEASURE

In the seventeenth and eighteenth centuries systems of measure were either oddly defined or simply not defined at all, and they were not relatable from one country to another, creating a mensurational morass. For example, the translator of Büsching's great geographical treatise found it necessary to add a thirteen-page section making conjectures concerning measures of length so that the variety of statistics could be reasonably compared.[43] The chaotic situation had been recognized as early as 1668 when Picard was asked to restore the standard French unit of length kept in Paris, the *toise du Châtelet*.[44] Through constant use the bar had become too long, but its origin was unknown and the length was undefined. Picard suggested using the length of a pendulum beating one second in Paris as a *rayon astronomique*. Similar proposals based upon the pendulum were made soon after by Mouton, by the vicar of Lyon, by Sir Christopher Wren in England, and by Huygens, the Dutch astronomer. Such a unit would be unchanging and based on nature, but the idea had a fatal defect: its application required that it be tied to one national place, and it would therefore be quite unacceptable to another nation.

In 1720 Jacques Cassini proposed that a "geometrical foot" should be one hundredth of one second of a meridian arc, and La Condamine suggested to the Académie in 1747 that an international unit of length be the length of a pendulum beating one second at the equator. He recognized the problem of national pride and proposed that foreign academies cooperate in its determination. Although the scientists of most nations were much in favor of a uniform system of measure, not much happened until the proposal made by C.-A. Prieur du Vernois.[45]

In 1790 Prieur, a military engineer, sent a memoir to the National Assembly concerning the need for a uniform system of measures with a fixed basis and a well-devised system of multiples and subdivisions. He proposed using a pendulum length. Tallyrand submitted Prieur's plan to the National Assembly and suggested that the British Parliament and the Royal Society be invited to cooperate. Similar proposals for a system of measures were made at about that time in the House of Commons and the United States House of Representatives. In Paris the Assembly asked the Académie to draw up proposals, and a committee recommended using a part of a meridian quadrant as the basic unit. The Assembly approved and in 1791 the Académie established five commissions to undertake various aspects of the work including the determination of the Dunkirk-Barcelona arc of the meridian. Even the increasing radicalism of the times did not stop the work, and the new National Convention was presented with a report in 1793 recommending, among other things, that a provisional meter of 443.44 Parisian *lignes* be accepted and that the various units of length, angles, and time should be subdivided decimally. During the next year or so the work on the new system was greatly curtailed as the National Convention dissolved the Académie, and two of the five commissions lost their leaders—one, Lavoisier, to the guillotine and one, Cordorcet, to poison to escape the same fate.

Fortunately Prieur kept his head, figuratively as well as literally, and when reason began to prevail in 1795 he urged that the work be again carried on. A detailed law was passed that same year, and the new system defining length, mass, and volume standards was introduced. After many difficulties in surveying the Spanish part of the arc, its length was established in 1799 in toises and the length of the earth's quadrant calculated. The definitive meter, one ten millionth part of the quadrant, turned out to be 443.296 lignes. The new system was not popular in France, and in 1812 Bonaparte decreed that the old measures could be used as well as the new. In 1816 Louis XVIII went further and prohibited the use of the new system. It was not until 1837 that a law was passed in France which abolished the old units as of 1 January 1840.

Meanwhile other countries began to adopt the metric system: the Netherlands in 1816, Greece in 1820, Spain in 1849, Italy and Serbia in 1863; and geodesists and surveyors often used metric units. Although the British pushed hard to have the yard, officially given a standard in 1855, replace the meter by distributing several dozen copies of the official yard to foreign governments, the meter forged ahead. The European Association of Geodesy adopted it in 1868, and the French organized an international metric convention in 1870 in Paris and again in 1872 which established a permanent committee. Finally, in 1889, using bars and weight standards cast in London, official metric standards were allocated to thirty member countries of the International Metric Convention.

In the course of two hundred years the confusing, archaic hodgepodge of length measures had been exchanged for a decimal system based on the earth itself. Although its development and introduction occurred at the time of the revolution in France and is usually thought of as being a child of it, it had its beginnings before and no doubt would have come about anyway. The growth of concern with science and measurement demanded it.

### The Growth of Interest in Social Affairs and Statistics

The development of a concern for science and for a universal system of measure from the mid-seventeenth to the mid-nineteenth century in Europe was accompanied by a parallel growth of interest and competence in statistics and statistical method, especially in physical science. In the realm of social affairs some of this was based on theoretical aspects, for example probability theory, some on problems resulting from growing populations and industrial growth, and some simply on the burgeoning interest in descriptive geography. For a considerable period in the two centuries the concern focused on such topics as population, fertility, mortality, and the comparative characteristics of nations. The development of statistical theory in social science and the mapping of statistical data took place later.

The social statistics of primary interest for some time were those having to do with population change. In the preindustrial period populations grew only slowly, with even that growth periodically interrupted by famines and epidemics, which were common and devastating. For example, bubonic plague took perhaps a sixth of London's population in the Black Death of 1664–65. Average life expectancy was only twenty to thirty years, and most of the people of western Europe were tied to the land in a precarious existence. The statistical study of population was initiated by John Graunt by his analyses of the death records of London parishes published in 1662. Several editions followed, one published by the Royal Society, and one, after Graunt's death, by Sir William Petty, who added estimates of economic consequences of deaths and created the term "political arithmetic."[46] This subject was pursued more aggressively in France, and especially by Johann Süssmilch in Germany during the succeeding century, where statistical bureaus had been established, but the development of the study of man was not nearly as rapid as the advances in physical science during the same period.

What we would call today social science and population studies require adequate statistics as well as the assumption that people's lot depends to a considerable degree on themselves. The development of a science of social affairs was greatly inhibited by the emphasis on the Greek classics and by the Cartesian ideas which permeated much Western intellectual thought until toward the end of the eighteenth century. Descartes's mechanistic physics and deterministic behavioral precepts left little room for empirical study.

The time lag between the development of the physical sciences and that of the social sciences is strikingly illustrated by thematic mapping. Numerous maps by students of the physical world, on such subjects as magnetic phenomena, currents, and geology, appeared from 1650 on. By contrast, so far as we know now, very few maps of social subjects, such as population, religion, or production, appeared before 1820. To be sure, as we will see in chapter 7, there were premap, graphic representations of statistics in the latter part of the eighteenth century by such notables as A. F. W. Crome and William Playfair, but there were no real thematic maps portraying statistical data.

In addition to the Cartesianism which retarded the growth of social science, another inhibiting factor was the lack of reliable numerical data. Although enumerations of various kinds are known from ancient days, they usually were partial and took little account of total populations and their characteristics. The modern era of censuses began in Europe. In 1686 Sweden made compulsory the parish record by the clergy of births, deaths, and marriages for all residents, and in 1670 Colbert extended the Parisian system of registering "domestic occurrences" to the rural communes. Not until the mid-eighteenth century, however, did censuses really get under way, and even then the results were far from satisfactory.[47] Sweden again led the way in 1749, followed by Finland (1750), Austria (1754), and Norway and Denmark (1769). The first French census was taken in 1801.

In England there was strong opposition to the enumeration of the people and those receiving parochial relief, but by 1800 attitudes had changed, perhaps owing to the growing population and concern for its subsistence, and an enumeration took place in 1801. The reaction was no doubt greatly influenced by the publication in 1776 of Adam Smith's *An Inquiry into the Nature and Causes of the Wealth of Nations,* and in 1798 of Malthus's first *An Essay on the Principle of Population . . . ,* the gloomy theory that, ceteris paribus, population growth would stabilize at the limit of subsistence and would be held there in an unhappy state by such forces as war and famine. These two works and Malthus's subsequent writings had much influence in Britain and abroad, helping to promote greatly expanded inquiry into social matters which required statistical data.

Most principal countries began their modern series of censuses between 1825 and 1860. By 1840 statistical offices had been established in many countries, and the exchange of ideas and techniques was common.

Quite apart from mapping, any series of numbers representing geographically based information presents numerous problems for the analyst, particularly, of course, those having to do with accuracy and completeness. A seemingly simple question is how to array such numbers in tables. The advantage of arranging them in columns and rows became apparent before 1750, the credit for doing this often going to a Dane, Ancherson. Geographical statistics related to countries and their subdivisions were later arranged not by individual areal units, but by categories of information in what came to be termed the Büsching method, after A. F. Büsching, called by V. John the "creator of modern geography" and "the founder of comparative statistics."[48]

Another aspect of statistics that influenced the intellectual developments during the period 1650–1850 was the development of probability theory. The idea that error is subject to regularity forced itself on researchers. From the time of Huygens in the mid-seventeenth century, the character of expectation and chance had been studied, and by the end of the eighteenth century the works of Laplace and Gauss had established the curve of error and the method of least squares. In 1812 and 1814 Laplace published first a comprehensive and then a popular discourse on probability in which he extended its usefulness from the natural sciences to the social sciences.[49] This was greatly extended in the 1830s by Adolphe Quetelet (or Quételet),[50] the royal astronomer of Belgium, who developed the concept of the "average man" and studied the moral attributes of man in such matters as crime and disease and the relation of these to education, age, and other characteristics. Similar statistical studies were carried out in France and England, and after the 1820s many statistical maps of moral characteristics were made. The concepts of regularity and relationship in statistics had been extended to include geographical occurrence.

John K. Wright characterized the period from 1650 to 1850 as the "age of geographical measurements."[51] Many of these reckonings were of the natural world, but they were all expressed in numbers which were subject to error and interpretation. For some time during that period statistics were collected willy-nilly, just for the gathering, so to speak, but it was not long before the gathering and manipulating of the numbers was done purposefully. There were immense difficulties: the systems and institutions for taking inventories of lands and peoples had to evolve; classicism and Cartesianism retarded the development of social science; and the understanding of the logical attributes of numerical data had to be worked out by the mathematicians. The studies of social affairs and statistics were tardy in getting under way during the two centuries; but they came on with a rush in the first half of the nineteenth, and the phenomena they dealt with were quickly mapped in a variety of ways.

### THE ORGANIZING OF INTELLECTUAL ACTIVITY

Before the mid-seventeenth century, the scholar-scientist was a rather unusual individual in his intellectual life. Although there were opportunities to discuss problems and exchange ideas, there were in fact very few such persons. Scientific societies first started in Italy in the sixteenth and seventeenth centuries, but they did not last long. Clerical opposition or outright suppression caused the last of the three that did form to end in 1657. In the seventeenth century leadership in natural science passed from Italy to England, France and Germany.[52] In the seventeenth century many scientific societies and groups were formed in France, both in Paris and in other cities. The Académie des Sciences was established in 1666 with an elite membership of only twenty. It was reorganized with an enlarged membership and a new constitution in 1700, and its scholarly publication, the *Mémoires,* first appeared in 1702. The Académie continued until the French Revolution, when reaction against its aristocratic character caused its suppression in 1793. It was replaced by a branch of the Institut National in 1795. The former name, the Académie des Sciences, was restored as one of the academies of the newly constituted Institut de France in 1816. Its *Comptes Rendus* began publication in 1835.

Like the Académie, the Royal Society of London began as an informal gathering of a few scientists. It first met in 1645, and in 1662 it received a royal charter. Its *Transactions* began publication in 1665. The reputation of the Académie and the Royal Society inspired the establishment of similar societies in other areas, and during the course of the eighteenth century most other Western countries established similar organizations.[53] Germany, no doubt because of its political fragmentation at the time, nurtured several. At Berlin an academy of sciences came into being about 1710, at Göttingen in 1751, at Mannheim in 1755, and at Munich in 1759.

Many of the members of scientific organizations were scholars whose interests were not necessarily restricted to natural phenomena. Philosophy and science were all-encompassing terms, and many of the societies were broadly based intellectually, such as the Royal Society in London, which had as members numerous physicians, as well as such political economists as Adam Smith and Malthus. Except for literary societies, which are as old as the scientific and philosophic societies, the development of scholarly organizations with more focused interests only began their rapid rise in the early nineteenth century. From the point of view of cartography, the most significant are the statistical societies, which reflected the burgeoning interest in political arithmetic, the structure of society, and man as a social being. Many were established throughout Europe and America, held regular meetings, and even had royal blessing, such as the Statistical Society of London.[54] In Britain alone such societies came into being during the 1830s at Bristol, Belfast, Glasgow, Leeds, Liverpool, and Manchester, to name several, but the majority died out by the 1840s.

Special note should be taken of the establishment of independent organizations devoted primarily to the exchange of scientific ideas in various fields and the provision of systematic direction to scientific inquiry. These were felt to be necessary because of the lack of such activity in the universities. This took place first in Germany, where the German Association was founded by Oken, a naturalist, in 1822, and which was considerably promoted by Humboldt. A similar organization, the British Association for the Advancement of Science, was founded in 1831 for like reasons. The regular meetings of its branches were attended by foreign scholars and did much to internationalize activity and exchange in all fields of investigation.

During the two centuries from 1650 to 1850, the organization of higher education varied a great deal from country to country and changed considerably as time went on. In England the universities emphasized classical themes almost to the exclusion of natural history, science, and current mathematical thought. In France the central university in Paris maintained control over all schools in every province. The development of more specialized schools, such as the Ecole des Ponts et Chaussées in 1747 and the celebrated Ecole Polytechnique in 1793, and numerous provincial academies during the eighteenth century, made France the leader in scientific education. In Germany, by the end of the eighteenth century, the universities had attained an eminent position, but generally the philosophy of scientific research was more inclusive and less exact than that of France. Many outstanding students of science in Germany lived outside universities and owed their allegiance to Paris as the intellectual center.[55]

The reputation of the Ecole Polytechnique spread rapidly, and polytechnic institutes were soon established elsewhere, such as at Berlin in 1799, Prague in 1806, and Bavaria and Vienna in 1815.[56] It is worth keeping in mind, however, that even though universities had grown considerably during the eighteenth century and the first part of the nineteenth, many of the important advances in science and technology—and in thematic cartography—were made by individuals who had little or no connection with universities, such as W. Smith (first geological map of Britain), Humboldt (first isothermal map), and Dupin (first moral statistics maps). Not until long after thematic cartography had matured were universities themselves to become the sources of thematic maps.

An important factor in the spread of knowledge during the eighteenth and nineteenth centuries, and a significant outlet for thematic maps, was the scholarly periodical. Other than a few exceptions such as the *Transactions* of the Royal Society, the *Mémoires* of the Académie and the Parisian *Journal des Savants,* the scholarly, scientific periodical began to flourish only after 1750. For example, only five such journals were initiated between 1700 and 1750, but by 1800 there were seventy-four, twenty-five of which were established in the decade 1790–99.[57] In the first half of the nineteenth century the rise in the number of periodicals devoted to separate fields became a flood. These covered a variety of subjects without clear boundaries and ranged over the whole gamut of natural history and science. No count of all the societies and their publications in western Europe has been made, but by 1850 they must have numbered in the hundreds. Each major city or province had its geological, statistical, philosophical, biological, chemical, and so on, society and journal. These offered a ready outlet for scholarly studies and thematic maps. For example, Humboldt published his innovative paper on isothermal lines in the *Mémoires de Physique et de Chimie de la Société de Arcueil,* and the map appeared in the *Annales de chimie et de physique,* both in 1817, and the former was translated into English in the *Edinburgh Philosophical Journal* in 1820 and 1821.[58] Similar examples of the circulation of ideas among the scholarly community of western Europe in the first part of the nineteenth century could be detailed endlessly. A century or so earlier science and natural history had just begun to escape from being parochial; by 1850 they had become truly international.

Another example of the spread of knowledge, both geographically and throughout the developing middle-class populations in western Europe, is provided by the design and trade in encyclopedias.[59] Up to the eighteenth century most encyclopedias had been clerical and historical and had, of course, appeared in Latin, which limited their use to the scholarly few. By the early eighteenth century alphabetical arrangements in the vernacular, aimed more at the arts and sciences, had been produced. In France, for example, the *Dictionnaire universal des arts et sciences* appeared in 1690, and *La Dictionnaire des arts et des sciences* was commissioned by the Académie Française for publication before 1700. In Britain Harris's *Lexicon Technicum* (1704) initiated the change which was taken much further by Chambers's *Cyclopaedia* in 1728 (no relation to the modern *Chambers's Encyclopaedia*). The latter was very well received, and several editions were issued in France. A translation was planned but instead turned into the controversial, and popular, *L'Encyclopédie* by Diderot and d'Alembert (1751–65). The upheaval in knowledge and outlook in the eighteenth century called for new approaches in coverage and organization, such as the *Larousse* in France and the *Encyclopaedia Britannica* in Scotland. The first edition of the latter was produced between 1768 and 1771, and new editions have appeared regularly ever since. Several encyclopedias appeared in Germany in the eighteenth century, to be followed by the innovative *Konversations-Lexikon* (1796–1811) of Brockhaus. Immediately thereafter *Der Grosse Brockhaus* was initiated, and it too has been issued ever since in numerous editions. In addition to these well-known encyclopedias of the late eighteenth and early nineteenth centuries, the almost insatiable market that had grown up saw numerous less successful attempts.

It is fitting that this survey of the organizing of intellectual activity from the mid-seventeenth to the mid-nineteenth century concludes with the encyclopedia. An encyclopedia is an enormous and expensive undertaking, issued in many volumes, and is dependent upon a wide sale for commercial success. The fact that so many could succeed, if not endure, by the end of the period indicates what a change had occurred in the intellectual development of western Europe. Knowledge had become organized and disseminated to a degree that a wide interest in all kinds of phenomena had developed. This was the soil in which thematic cartography could grow.

## The Growing Interest in the Environment

Attitudes toward the natural environment underwent a complete turnabout during the two centuries from the late 1600s to the late 1800s. This remarkable change occurred because of a series of complex influences, and no attempt to describe them all is possible in a few paragraphs. Nevertheless, a review of the major factors is warranted because they had a marked effect on the emerging field of thematic cartography. The two most significant are first, the attitudes generally grouped under the umbrella term "romanticism," and, second, the growing concern with the conditions of the natural environment and their relations with health and the quality of life. Of the two, romanticism is the more diffuse and elusive.

Up to the early part of the eighteenth century the dominant philosophical concepts were the classical notion of an orderly ideal, reinforced by the idea that anything disorganized reflected a lack of divine grace. For example, the study of the disorganized landform would be fruitless, for it would not lead to any order. This began to change in the seventeenth century, initiated by such writers as Burnet in England and, later, Buffon in France.[60] Both wrote on the origins of the earth, and their works enjoyed great popularity, although they were deeply unappreciated by the clerically inclined. Buffon, for example, was the first to bring the element of time into the picture. The new attitudes spread to art, poetry, and literature, and nature began to be appreciated with a kind of reverence.

As we have seen earlier, the systematic study of the earth led to steady developments in mineralogy and geology. Other aspects of natural history came under scrutiny, such as the biological world. In the late seventeenth century the fascinating microscopic protozoan life became known through the work of the Dutchman Leeuwenhoek, and the botanists and zoologists were busy describing and trying to classify plants and animals. Comparative studies were initiated, exemplified by Cuvier's work in the late eighteenth and early nineteenth centuries, which essentially founded the science of paleontology. The scientific study of nature was off and running.

As we saw earlier, concern with social conditions was also gaining during this period, partly owing to the great increase in population in some areas and not in others, the rapid growth of cities as a result of the industrial revolution, and the breakdown of the traditional agricultural systems. Like the natural environment, man, his conditions, and his character became legitimate areas of study. And, also like the natural environment, aspects of the social environment often needed mapping to bring distributions into focus and to study geographical relationships.

An interesting instance of a combined concern for the social and natural environment, which resulted in considerable significant thematic mapping, is that having to do with medical geography in general and maps of disease in particular. The relation of environmental factors to health, or the "medical topography," as they were often collectively referred to earlier, has been of interest ever since the work *Airs, Waters, and Places,* presumably written by Hippocrates himself. A systematic study of medical geography, without maps, appeared in 1792,[61] but it was not until after the first pandemic of cholera began in India in 1817 and reached Europe a dozen or so years later that investigations were initiated into the significance of environmental factors on the incidence of the disease.[62] A great many interesting and innovative maps were prepared by medical practitioners in the hope of identifying environmental factors which might influence the progress and severity of the disease. These "sanitary maps" of the environment form an unusually fascinating chapter in the development of thematic cartography, perhaps because many of them were made for analytical purposes as well as for display.

### Industrial and Commercial Development

The change in scientific knowledge and interests and the expansion of man's outlook to encompass all aspects of the environment during the period from the mid-seventeenth to the mid-nineteenth century amounted to a complete intellectual transformation. Paralleling that change, and fully as remarkable, are the developments that took place in technology, industry, trade, and transportation. These too gave rise to new kinds of maps which fall in the thematic class.

Every aspect of life changed during the eighteenth century. In agriculture the breakdown of the open-field system, the introduction of the modern plow, crop rotation, improved tillage practices, and livestock breeding raised yields enough to compensate for the decrease in farm labor that resulted from the migration to the growing urban centers. At the same time, fundamental changes were under way in manufacturing, mining, and smelting.

What we now call the industrial revolution began in Britain and soon spread to the Continent. It started in textile manufacturing toward the mid-eighteenth century with the invention of the flying shuttle in 1738, which doubled the production of the hand weaver. Leapfrogging inventions followed: advances in spinning to keep up with the weavers, advances in weaving to keep up with the spinners, and so on. The steady series of inventions, mostly by practical men, not theorists, caused enormous changes. Eloquent testimony is provided by the simple fact that the importation of cotton to Britain from the West Indies and America increased from some 7,000,000 pounds in 1780 to

32,000,000 by the beginning of the French Revolution only a decade later.

Developments in mining and smelting were just as remarkable. Until nearly the mid-eighteenth century charcoal was required to smelt iron. Forests were badly depleted, and this, together with the primitive method of refining the raw iron into pigs on forges, severely retarded the mining industry and the manufacturing of iron and steel. The discovery that coke could be used in place of charcoal and that the iron could be refined in a furnace gave further impetus to the industry, and it grew rapidly. For example, in Britain during the eighteenth century, the production of iron multiplied thirty or forty times.[63]

The demand for coal, for smelting and for heating, called for more mining, and this necessitated some way of removing groundwater from the deeper and deeper mines. Newcomen's steam pump of the early eighteenth century was perfected by Watt after midcentury, and before the end of the century he had improved it until it could drive a wheel and thus turn machinery, such as the spinning mule and, later, the power loom. The transformation was contagious in other fields such as metalworking and machine building, but especially so in the chemical industry, closely allied to textile manufacture because of the need to wash and bleach materials. The pace of invention quickened as the eighteenth century wore on: in Britain alone in the successive four quarters of that century the numbers of patents issued were 116, 176, 458, and 1349.[64]

These momentous changes brought about social changes of similar magnitude. Shifts of population, urban crowding, enormously increased pollution, deterioration in living conditions, and technical demands calling for changes in education were some of the immediate consequences. Such ramifications, calling for remedial action, promoted the need for censuses and surveys and gave impetus to the growth of the fields of statistics and what the Belgian statistician Quetelet called social physics, the study of the "average man" and variations from the norm, which have already been mentioned. Although the changes that took place in agriculture, manufacturing, mining, and other economic activities directly evoked very few thematic maps during the eighteenth century, the consequences that followed did. Some of these have already been mentioned, such as the sanitary maps and moral statistics maps. Maps of population and manufacturing also came along in due course. One area of development that stemmed from the industrial and commercial growth of the eighteenth century—transportation—did result in some new kinds of thematic maps.

The increasing production of the growing industries, the need for transporting greater amounts of the raw materials and finished products, and the general urge toward greater mobility put heavy demands on a thoroughly inadequate road and canal system. The roads were particularly bad during much of the eighteenth century. They tended to be better on the Continent than in Britain, where, except for a very few stretches of turnpike, they were well-nigh impassable most of the time. Carriage of goods was usually by pack horses, and during the wet winter season roads were hardly usable at all. It is said that in Cornwall wheeled carts were scarcely used before 1800. Conditions were somewhat better in France, where the Corps des Ingénieurs des Ponts et Chaussées had been established in 1716. Advances in the technique of road-building in the latter half of the eighteenth century led to steady improvement of the main roads everywhere, but even by 1830 the average speed of the mail coaches in Britain, much the fastest mode of wheeled transport, was only nine miles an hour.

During the period from the seventeenth century until well into the first half of the nineteenth, the only alternative to inland transport by road was by waterway. Some rivers were conveniently located and had the necessary depth, but many had to be canalized, and more often canals had to be excavated and locks constructed. The plains of northern Germany, the Low Countries, and northeastern France all benefited early from a system of connecting canals and rivers. By the mid-eighteenth century an extensive network existed on the Continent, but in Britain the canal era was only beginning at that time, the first being the seven-mile canal used for taking coal from Worsley to Manchester. Activity was feverish thereafter, and by 1800, even in Britain, the waterways were quicker, cheaper, and safer than the roads. They carried the bulk of raw materials and finished goods and were an indispensable adjunct to the furtherance of the industrial revolution.

The adaptation of the steam engine to turn a wheel had a remarkable effect on transportation. Horse-drawn wagons riding on wooden rails were used even before 1650 in Germany in mining operations, and their use by mine operators spread throughout western Europe. In the eighteenth century the wooden rails were replaced by iron. The adaptation of the steam engine to turn a wheel in the 1780s provided the basic technology for a self-propelled mechanism, and the first practical locomotive was built in 1812. Steady progress was made, and between 1825 and 1835 steam-propelled passenger and freight carrying railways were put into use in all the countries of western Europe. The growth in mileage of the railway network was phenomenal, as shown by the accompanying table.

APPROXIMATE TOTAL MILES OF PUBLIC STEAM RAILWAY LINES IN GREAT BRITAIN, FRANCE, AND GERMANY

| Year | Miles |
|---|---|
| 1820 | 0 |
| 1830 | 70 |
| 1840 | 2,075 |
| 1850 | 11,670 |
| 1860 | 22,100 |
| 1870 | 36,600 |

(From *Chambers's Encyclopaedia* and other sources.)

The development of the steam-powered seagoing vessel in the early 1800s and the iron hull before midcentury began a similar fundamental change in ocean transport. The movement of goods and passengers on all forms of transportation grew phenomenally, as did the development of statistical thematic maps to display the geographical structure of the flows.

EPITOME

Plato's aphorism that necessity is the mother of invention is usually invoked to account for the fabrication of some contrivance, and not ordinarily for the development of an entirely new field of endeavor. Yet it fits thematic cartography as well as anything else. The fact that the necessity arose because of the remarkable series of intellectual, social, and economic developments which took place in western Europe over a period of some two hundred years makes no difference. The changes that occurred were such that a new kind of map was needed. It required new attitudes toward the ancient art of mapmaking, new data, new symbolism, and new techniques of duplication. When viewed against the general field of mapmaking, thematic cartography was indeed an invention.

# *Three*

# FROM SINGLE MAPS
# TO ATLASES

The growth of thematic cartography was not a sudden and spectacular development. Although there were some remarkable individuals who contributed a good deal in its very early stages, such as Athanasius Kircher, and especially Edmond Halley, in general its growth from the latter half of the seventeenth century through its first hundred years was relatively slow. The pace picked up toward the end of the eighteenth century when a larger variety of thematic maps began to appear, but the new type of map really showed its versatility in content and method only during the first half of the nineteenth century. In this chapter we shall try to survey this uneven development from its meager beginnings through its period of acceleration and maturation, so that the closer look at the classes of innovative maps to be dealt with in the succeeding chapters will have a firm chronological context.

The new approach to mapmaking did not in any way burst on the scene: the initial examples are few, they are generally primitive in symbolization (though not always), and they reflect a lack of data. The few early, truly thematic maps of which we now know, with few exceptions, either are manuscript maps or were included as illustrations in treatises. It is very likely that there are undiscovered manuscript thematic maps in various archives that were prepared as addenda to documents that were never printed. They will probably remain largely undiscovered, since, on the one hand, historians of cartography are not likely to be consulting such documents, and on the other hand the scholars who do are not likely to realize the significance of such maps to the history of cartography.

A happy example of an exception to the foregoing is provided by Dainville, who describes four sheets of a map of Protestant places in France in 1620, found among the manuscripts of the Bibliothèque

*From Single Maps to Atlases*

Municipale de Grenoble.[65] The maps, made about 1640, show by colored symbols the Huguenot strongholds in France during the Protestant-Catholic struggles of the early seventeenth century. Another example is a pre-1724 manuscript map reported by Engelmann which showed the customary beverages (wine, beer, etc.) of German and European regions.[66] There must be scores of similar maps attached to the surveys and reports in government and other institutional archives. They will come to light only gradually, and, because our knowledge of them is so scanty, I will restrict my attention, in most cases, to printed maps.

One notable fact about the scientific, scholarly activity in the seventeenth and eighteenth centuries, as compared with its character later on, is that the same men actively studied all the sciences. There was no separation of physics, chemistry, astronomy, or mathematics, and the same names appear in every field. For example, the earliest-published truly thematic map now known is by Athanasius Kircher. His career exemplifies how scientifically cosmopolitan he and others like him were.

Kircher was born in 1602 in what is now the Federal Republic of Germany, and he excelled in geography at elementary school. At the age of twenty-six he was called to teach mathematics, moral philosophy, Hebrew, and Syrian at the Jesuit University in Würzburg. He fled to escape the invasion by Gustav of Sweden, and for a time he was at Avignon, where he concerned himself with mapmaking, mathematics, and the study of hieroglyphics.[67] In 1634, the year after Galileo was forced into seclusion, Kircher became professor of mathematics at Rome, where he remained until his death in 1680. Among other things, he invented the magic lantern and experimented with an early form of the microscope. Apparently a tireless scholar, he wrote more than forty volumes.

Kircher was much impressed by the Calabrian earthquake of 1636. The tremendous forces locked within the earth, as evidenced by volcanoes, kindled his desire to delve into the nature of the interior of the earth and its relation to surface phenomena. He carried on an extensive correspondence with members of the Jesuit order stationed around the world, and in 1664 he produced one of his better-known works, *Mundus Subterraneus*.[68] That work, published first in Amsterdam in 1665 and then in two subsequent editions, contains a number of diagrams concerning the circulation of water in channels beneath the earth's surface and the origins of rivers, all quite fanciful; but it also contains thematic maps of the horizontal circulation of ocean waters, including one of the world entitled "Tabula geographico-hydrographica motus oceani currentes..." (see FIG. 29), as well as one of the Mediterranean and another of the Americas. Kircher actually holds a more important place in the history of cartography for a thematic map that he never published but that he wrote about. In his *Magnes sive de arte magnetica opus tripartitum,* published in Rome in 1641, Kircher described how to make a map showing the distribution of magnetic declination over the earth.[69] In the third edition of *Magnes sive* (Rome, 1654) Kircher states that he could have produced such a map with the observations he had collected and that he would have done so if the cost and the printing schedule had permitted it.[70] I will refer again to Kircher's concern for mapping declination.

Five years after Kircher's death in 1680, another map of ocean circulations and tides was produced by Eberhard Werner Happel (Happelius), entitled "Die Ebbe und Fluth auf einer Flächen Landt-Karten fürgestellt," a portion of which is shown in FIGURE 11. It shows the movements in the same way as Kircher's map, with a kind of streamlining with no indication of direction.[71] Happel is another example of the versatility of the time: he was also a novelist.

The first published thematic maps appear to have been concerned with portraying the structure or pattern of a dynamic geographical phenomenon, the circulation of ocean waters. This is not surprising since wind-powered, oceanic navigation was a very chancy activity in the seventeenth century, and all natural phenomena which affected its speed and safety were matters of great concern. Although a general map of ocean currents was thought to be very useful, by Samuel Pepys among others, the maps of Kircher and Happel were crude and could not be much bettered without great increases in data.[72] One phenomenon, much easier to observe than a current, is the direction of the wind, and since it was already known that the main ocean movements seemed to follow the wind, it is to be expected that atmospheric circulation would be a subject for an early thematic map. One such was published in 1686 by Edmond Halley as an illustration to accompany a paper describing and attempting to account for the earth's wind systems, a portion of which is shown in FIGURE 12.[73] FIGURE 21 shows the entire map.

### Edmond Halley as a Thematic Cartographer

Edmond Halley is clearly the first thematic mapmaker who merits the name cartographer, and he was a most illustrious one.[74] He was born in 1656 and is popularly remembered for the comet which bears his name. His forecast that the comet would reappear about seventy-five years after it was observed in 1682 turned out to be correct, and the prediction had a profound effect on allaying popular fear and superstition about the dire influences of these celestial displays. Halley was educated at Oxford and became a member of the Royal Society in 1678. He was editor of its *Philosophical Transactions* from 1685 to 1692. In

**Figure 11** A portion of the western section of Happel's 1685 "Die Ebbe und Fluth...." Reduced 40 percent. Copperplate engraving. (Courtesy of the British Library.)

1704 he was named Savilian Professor of Geometry at Oxford, and in 1720 he succeeded Flamsteed as the astronomer royal. He led a most active life, ranging from holding the rank of captain in the Royal Navy while commanding voyages of scientific investigation to carrying out numerous studies resulting in original contributions to several branches of astronomy, physical geography, and geophysics.[75]

The wind chart is the first of its kind as well as being Halley's first thematic map. Since no traditional symbolism for showing wind direction and strength existed, Halley had to devise one. He appreciated the problem of trying to make clear the complex structure of the wind systems at the earth's surface and flatly stated that it could be done better by a map than by words. He explained his symbolism as follows:

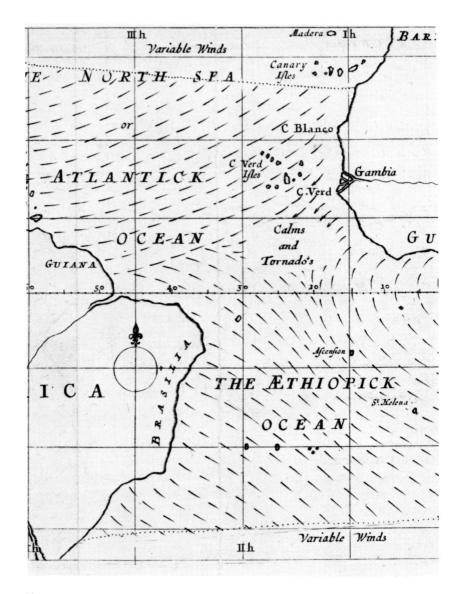

**Figure 12** The Atlantic portion of Halley's 1686 map of the trade winds and monsoons. Nearly full size. Copperplate engraving. (Courtesy of the British Library.)

*From Single Maps to Atlases*

> The limits of these several Tracts are defined everywhere ... by drawing rows of stroaks in the same line that a ship would move going alwaies before it; the sharp end of each little stroak pointing out that part of the Horizon, from whence the wind continually comes.

After Halley's map most maps of winds used arrows to show wind directions, with the arrowheads pointing downwind. Curiously, this was done on Halley's map only in the vicinity of Cape Verde (FIG. 12).

Halley's wind system was frequently portrayed on many English and Dutch charts in the first third of the eighteenth century, and the pattern of winds appeared regularly on maps from the mid-eighteenth century onward. They employed a variety of symbolism, usually arrows, sometimes cherub's heads puffing, or simply named zones, but Dainville reproduced a 1737 map showing the winds of Languedoc which employs five kinds of symbols.[76]

Edmond Halley merits the honor of being known as the first versatile thematic cartographer for two reasons: (1) he was the author of four clearly thematic maps and one quasi-thematic one, and (2) he was the first to publish a map employing and "popularizing" isolines as a means for showing the spatial variation (structure) of a phenomenon, a technique soon to be adopted by others. If Halley had not done so, someone else probably would have; nevertheless, he was the first. The two isoline maps are perhaps his most well-known contributions, since they attempted to show the distribution of compass variation, a subject of great concern. They will be discussed in detail in the next chapter.

Halley's other two maps are of widely divergent subjects which show his versatility and illustrate the wide-ranging interests of the scientists of the time. One of these is less clearly thematic than his other maps, this being "A new and correct Chart of the Channel between England and France ... showing the sands, schools, depths of Water ... with ye flowing of the Tydes and the setting of the Current...."[77] It was an ingenious cartographic accomplishment employing roman numerals to show the times of high water, which together with data concerning the positions of the moon enabled one to calculate how long the tide would run eastward. Arrows, which Halley called "darts," showed the direction of the tidal set. It was the basis for maps used in *Great Britain's Coasting Pilot* for several decades.

The last map by Halley is an ingenious and very effective presentation of the passage of the shadow of the moon across England during a total eclipse of the sun in 1715 (FIG. 13). Its simplicity is deceiving, for it clearly indicates by graphic means the varying duration of totality and gives the timing of the eclipse in minutes past nine at London by numerals along the axis of the path. The use of the shading shows how fertile and imaginative was Halley's grasp of the potentialities of graphic portrayal. The dark ellipselike figure representing totality was to "slide" along the shaded path from southwest to northeast, and the relative duration of totality for any place along the path was shown by the width of the ellipse in line with that place. London, being close to the edge, "tis dubious whether it will be a Total Eclipse or no, London being so near ye southern limit." The map was "dispersed all over the

**Figure 13** Halley's map of the path of the eclipse of the sun in 1715. Original 225 × 290 mm. Copperplate engraving. (Courtesy of the British Library.)

*From Single Maps to Atlases*  51

Kingdom" as the main part of a broadside or leaflet before the eclipse, and it certainly is the first thematic map to have had wide popular distribution.

INDIVIDUAL EIGHTEENTH-CENTURY THEMATIC MAPS

The map of declination (compass variation) by Halley, along with his map of the winds, were in truth, as Chapman put it, "the parents of a most numerous progeny." As mentioned earlier, sea travel, when one was at the mercy of the wind and currents, was hazardous at best in the seventeenth and eighteenth centuries. Part of the danger stemmed from not being able to tell where one was, since the longitude could not be reckoned at sea by direct observation as could the latitude.[78] Various plans were proposed, some quite fanciful, and interest ran high, since the British Parliament had offered a generous prize for "finding the longitude." For more than a century the scheme of using declination had been thought workable, and in any case the compass was a most valuable navigational aid.[79] It is to be expected that any information about the structure of the magnetic field of the earth would be of great interest, particularly its relation to the direction and the inclination of the needle.

During the eighteenth century several dozen thematic maps of magnetic phenomena were made by a variety of individual investigators, and they employed the Halleyan curve lines of equal value to display the variations of whatever aspect they mapped. Hellmann terms the period from 1700 until the early nineteenth century as one of compilations of single observations as opposed to any plan of coordinated survey.[80] Among those making significant maps was Wm. Whiston, who in 1719–20 made the earliest map of equal magnetic inclination. Lucasian Professor of Mathematics at Cambridge, he reasoned that the lines of equal dip could not be more irregular than the Halleyan lines of declination, and, the "mutation" (secular change) being less, the lines of equal dip would be more useful to "discover the longitude." A variety of similar maps of particular areas were made, and by 1768 J. C. Wilcke of Sweden had made an inclination map of the entire world.

It is probably a safe observation that thematic maps of magnetic phenomena made during the eighteenth century far outnumbered other thematic maps. They were scientific and in no way glamorous, but they did do one thing for thematic cartography: they made commonplace, in scientific circles, the curve line, the Halleyan line of equal value, which was to become one of the primary means of representing quantitative distributions on thematic maps.

As early as the latter part of the seventeenth century, attention had been called to the desirability of mapping the character of the land. Martin Lister was an active fellow of the Royal Society and probably is best known as a natural historian for thoroughly describing present and fossil shells while refusing to his death to believe that they ever derived from living creatures. In 1684 Lister, commenting in the *Philosophical Transactions* of the Royal Society, pointed out the need for careful examination of the earth and wrote:

> And for this purpose it were advisable that a *Soil* or *Mineral Map,* [i.e., geological map] as I may call it, were devised.... The *Soil* might either be coloured, by variety of *Lines,* or *Etchings*... and the limits of each Soil appearing on the *Map,* something more might be comprehended from the whole... which would make such a labour well worth the pains.[81]

Students of the land were slow to do what Lister had suggested.

What has been called the first geologic map was made by Christopher Packe, a physician who practiced in Canterbury from 1726 until he died in 1749.[82] A preliminary map was prepared in 1737, and in 1743 he published the completed map, which he called a *New Philosophico-Chorographical Chart of East Kent...* (FIG. 14). The map is relatively large scale (ca. 1:40,000) for a thematic map, but Packe was primarily concerned with the interrelationships among streams, hills, and valleys, which he analogized, using medical terminology, to a physiological circulatory system. The valleys are darkened by a kind of vascular hatching which lightens upslope, leaving the ridges white. Arable lands, marshes, meadows, and beaches are symbolized, and spot heights are used. Packe did not intend his map to be a general map, but rather a "philosophical," that is, scientific, description based on the traditional macrocosm/microcosm analogy in which the character of the universe was thought to be reflected in the human body.[83] Packe exemplifies the versatility of the intellectuals of the time, and joins a surprisingly long list of physicians who have contributed much to the development of cartography.[84]

Except for Packe, Lister's proposal for what would be called geognostic (geologic) mapping was not followed up in Britain until nearly a century later. Geognostic mapping had an earlier start in France with the individual and collaborative work of Jean-Etienne Guettard and Philippe Buache. The former was a pioneer geologist and mineralogist who has been called the father of geologic maps, since he stressed the importance of mapping in the study of rocks and minerals.[85]

Guettard, born in 1715, was an ardent scientist and, like others mentioned above, he became a doctor of medicine.[86] He traveled a great deal, as part of the retinue of the duc d'Orléans, and spent much time in botanical observation. He observed that the distribution of plants seemed often to be dependent on the occurrence of certain minerals and rocks, and this phenomenon led him to the geographical study of mineralogy. In 1746 Guettard presented a *mémoire* to the Académie

From Single Maps to Atlases 53

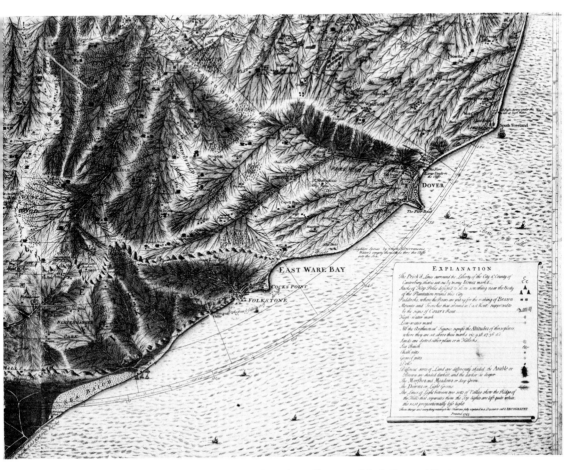

**Figure 14** A portion of Packe's 1743 "New Philosophico-Chorographical Chart of East Kent...." Original ca. 1:40,000, reduced 75 percent. Copperplate engraving. (Courtesy of the British Library.)

accompanied by two *cartes minéralogiques* prepared by Buache.[87] These two maps appear to be the first to show by dotted lines and shading the distribution of three separate and distinct *bandes* of rocks and minerals, a sandy zone, a marly zone, and a metalliferous zone. They are thus clearly thematic in that they attempt to show the spatial organization of a geographical distribution. In addition, each map employs nearly fifty symbols to show the occurrences of various rocks and minerals. The first map extends from Iceland and Norway to France, while the second is larger scale and is an enlargement of the center section of the first. It shows northern France and southern England to illustrate Guettard's hypothesis that the *bandes* continued beneath the Channel and reappeared in Britain. Three other similar maps showing the three *bandes* and the occurrence of minerals and rocks were made by Buache to illustrate other *mémoires* by Guettard, one of Switzerland, one of Egypt and the eastern Mediterranean, and one of northeastern North America.[88] As a geologist Guettard is best known as a pioneer paleontologist and as the first to recognize the existence of old volcanoes in central France, but his singular contribution to the development of thematic mapping is the portrayal of zones of similar composition, the forerunner of the standard geological maps. Guettard knew what he was doing; at the beginning of his 1746 *mémoire* he stated:

> If nothing can contribute more towards the formation of a physical and general theory of the earth than the multiplication of observations among the different kinds of rocks and minerals, certainly nothing can make us more aware of the utility of such research than to collect into one view those various observations by the construction of mineralogical maps.

Guettard's mapmaker, Buache, also a member of the Académie, made several thematic maps of his own. One of these is an innovative map of the Channel showing isobaths, first presented to the Académie in 1737. We will look more closely at the maps made by Buache in the next chapter.

A few other geologic maps were made in Europe, notably in Germany by G. C. Füchsel in 1762, and the first (hand) colored geological map by F. G. Gläser in 1775.[89] Generally, however, good quality, detailed, thematic maps of geology would have to wait until the nineteenth century.

Although greatly outnumbered by thematic maps made by students of natural history, a few maps of other subjects were prepared during the eighteenth century. Their symbolism is relatively primitive. One such series of four maps is by Gottfried Hensel, published in a book which appeared in 1741 in Nürnberg.[90] Perhaps inspired by Leibniz, who proposed the use of maps on which language areas could be marked, samples of the written language of each peoples are placed in proper geographical position. Dotted lines separate the language areas. FIGURE 15 is "Europa Polyglotta . . . ." Hensel's map of Africa (considered later) uses color to show locations of the descendents of Shem, Ham, and Japheth. His maps may be the first to use color to distinguish areas on a thematic map. The subject of language distribution and origin continued to interest philosophers, and nearly a hundred years later, in 1823, Julius Klaproth published his *Asia Polyglotta . . .* accompanied by a language atlas, consisting primarily of tables of equivalents of German words in Asian tongues and including one map.[91]

One quasi-thematic map on an economic subject was published during the eighteenth century by A. F. W. Crome in 1782 (FIG. 16). It is

**Figure 15** Hensel's 1741 "Europa Polyglotta. Linguarum Genealogiam exhibens...." Original 193 × 155 mm. Copperplate engraving, hand color. (Courtesy of the British Library.)

generally considered to be the first of its kind.[92] Crome, a teacher of geography and history at Dessau at the time, became professor of statistics and political economy (Kameralwissenschaft) at Giessen. His "Neue Carte von Europa welche die merkwürdigsten Producte... enthalt," published in Dessau, contains a variety of symbols to show the occurrence of fifty-six commodities, along with others to show cities, ports, and such. The political boundaries are enhanced with color. Crome's map was very well received and was issued in several editions during the two decades following its first publication. Its variety of point symbols which simply show geographical occurrence is reminiscent of Guettard's *cartes minéralogiques,* and in that sense Crome's map is as much a general or reference map as it is thematic.

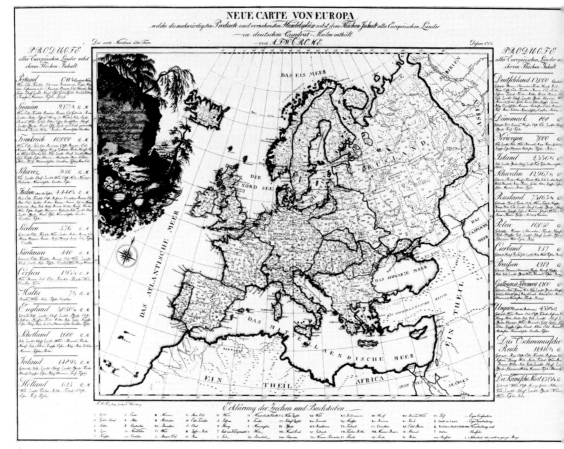

**Figure 16** Crome's 1782 "Neue Carte von Europa...." Original 705 × 545 mm. Copperplate engraving, hand color. (Courtesy of the Geografisch Instituut, Rijksuniversiteit Utrecht, Ackersdijck Collection.)

## Survey of Early Nineteenth-Century Mapping of Natural Phenomena

The beginning of the nineteenth century, the period around 1800, is a turning point in the history of thematic cartography for two reasons. As we have seen, during the eighteenth century, thematic maps were comparatively rare and unusual, such as Hensel's language maps, Guettard's mineralogical maps, and the mariner's utilitarian maps of compass declination. A few years into the 1800s thematic maps had ceased to be oddities, and by midcentury they had become commonplace. Their long period of adolescence had ended, and by the 1860s or 1870s they had acquired all the characteristics of maturity.

In the general history of cartography the move into the nineteenth century is notable for a technical development which was to have far-reaching effects on the field. In 1798 Aloys Senefelder, an aspiring actor and playwright in Munich, devised a printing process called lithography by which images could be reproduced from a printing surface much easier to prepare than those required by the then-standard processes of copperplate engraving, woodcut, and wood engraving.[93] The discovery of lithography, a perfect example of serendipity, not only opened the way for relatively inexpensive reproduction, but eased the problem of obtaining tonal gradation, an important technique in thematic mapping.

This is not the place to detail the exceptional effects lithography had on the emerging field (see chap. 7), but they were profound indeed. Suffice it to say that the new technique spread rapidly and after 1820 made possible the preparation of far more maps than likely would have been possible had thematic mapping been restricted to employing copper and wood as printing media. At the time the effect of lithography in the field of duplication was as great as has been the introduction in our time of rapid-copying techniques such as Xerox.[94]

As we have seen, the slow development of thematic mapping during the eighteenth century was primarily concerned with the physical environment, principally magnetic phenomena and wind and water currents of concern to navigation, and to a lesser extent with geological subject matter. This pattern continued into the nineteenth century, with the atmospheric characteristics of temperature and the distribution of vegetation being added to the repertoire.

In the eighteenth century the mapping of magnetic dip or inclination by Whiston and Wilcke had been added to the often reproduced and revised Halleyan maps of compass declination. The third element, the varying intensity of the geomagnetic field, was added first by Humboldt in a diagram in 1804 and then in a proper map of part of northwestern Europe by Hansteen in 1825.[95] In review, one could say that there was a steady growth in knowledge about and in the mapping of geomagnetic phenomena from as far back as the latter half of the seventeenth century. By contrast, geologic mapping, the bare beginnings of which took place in the eighteenth century, accelerated much more rapidly in the early nineteenth. This notable spurt occurred for two main reasons. The first was practical. The canal building era got under way in the eighteenth century, being slower in Britain than on the Continent, but by 1800 it was in full swing. The engineers needed maps of structures and rock types to plan the routes and to maintain the water supplies. Drainage of wet lands and the mining of coal and other minerals placed similar demands on the engineers.

This practical requirement was intertwined with an equally powerful interest of a purely scientific nature. The mystery of earth structures yielded to the observation that the ages of rocks could better be ascertained by analyses of fossil assemblages than by their mineralogical composition, based on Werner's fourfold primitive, transition, stratified, and alluvial succession. Noted first in England by the surveyor-engineer William Smith in the 1790s, it was extended in France by the work of the zoologist Georges Cuvier and the geologist-mineralogist Alexandre Brongniart, who worked out the faunal sequence of the strata of the Paris Basin in the first decade of the new century. The subject was one of intense scientific interest for it involved a direct challenge to the prevailing Wernerian doctrine of a primeval ocean. Furthermore, it brought into serious doubt the limited six-thousand-year age of the earth as interpreted from the Scriptures, and it gave assistance to those who supported the concept of catastrophism—that the earth had evolved through a series of sudden great changes, such as upheavals and floods, which periodically had destroyed life and changed the landscape.

This early work in geology resulted in two notable maps which became springboards for the rapid developments to follow. One was a geological map of the Paris Basin by Cuvier and Brongniart published in Paris in 1811, and the other was the first modern geological map of a large area by William Smith, published in London in 1815. Brongniart, with Cuvier as an associate, studied the structures around Paris in the early years of the century, published a first report in 1808, reported to the Institute in 1810, and issued a separate work with a map, sections, and drawings of fossils in 1811 (FIG. 17).[96] William Smith, born of a farm family in 1769 in Oxfordshire, had only a village school education, but even as a child he collected fossils. He became an assistant to a surveyor and in a few years was involved in canal surveying. He was very active and traveled all over Britain in connection with drainage, canal, and irrigation problems. He began work on his geological map in 1805, for which William Cary engraved by 1812 a new, large, fifteen-sheet, topographical base map which assembled measured 2.6 meters high by 1.9 meters wide. Even with its large scale, five miles to the inch, Smith's map can be classed a thematic as well as a general map. He sacrificed all his earnings for its preparation, but ultimately he became known as the father of British geology, receiving a medal from the Geological Society, a pension from the government, and an LL.D. from Trinity College, Dublin. In many ways a complete opposite of Edmond Halley in education and science, his contribution to the development of thematic cartography was also to become a milestone.[97]

The works of Smith, Cuvier, and Brongniart, and after them d'Halloy, not only established the principles of paleontological stratigraphy, they also improved the cartography involved in displaying the character of

From Single Maps to Atlases

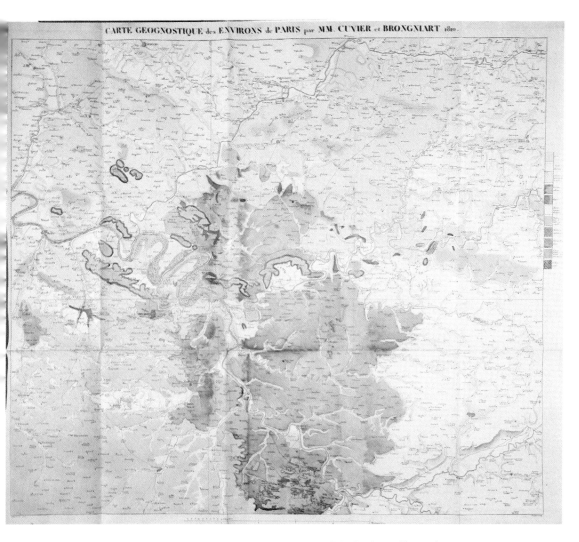

**Figure 17** "Carte Géognostique des Environs de Paris par MM. Cuvier et Brongniart 1810." Original 465 × 400 mm. Copperplate engraving, hand color. (Courtesy of the University of Wisconsin Memorial Library.)

the subsurface and formed the pattern for the increasing numbers of geological maps to follow. Whereas the maps of Guettard and Buache showed simply the horizontal distribution of surface materials, those of the nineteenth century were of a higher order, attempting to display structures. For example, Smith employed darker tones of color at the base of each formation, and the cartographers often included sections to help invoke in the map user an image of the three dimensions involved (FIG. 18).

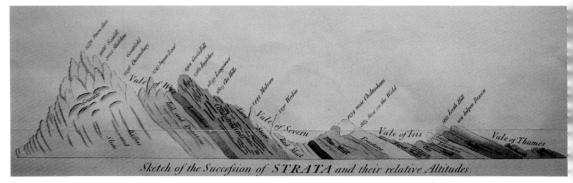

**Figure 18** Profile and section from W. Smith's 1815 "A Delineation of the Strata of England and Wales...." Copperplate engraving, hand color. (Courtesy of the British Library.)

In this general review of the progress of thematic cartography, I have focused initially on the geological developments, not only because they had earlier roots, but also because of the three-dimensional problems involved. Nevertheless, it should be emphasized that equally significant steps were being taken in two other areas of physical geography. Most notable from the cartographic points of view was the concept of the isotherm, borrowed from the Halleyan lines by Humboldt to show the distribution of temperature. First described in 1816, the famous diagram-map was published in curious circumstances in 1817.[98] Humboldt's map had great impact not only because of the fundamental importance of the isoline technique as a means of portraying the variations of the particular phenomenon, but probably also because of its author's scientific renown. It has been said that Alexander von Humboldt "became next to Napoleon Bonaparte the most famous man in Europe [and that fact] did more than anything else to raise the natural sciences in the popular mind to that eminence which earlier belonged to polite literature."[99] The concept of the isoline soon was employed in mapping other meteorological phenomena, such as atmospheric pressure, precipitation, numbers of stormy days, and so on, and by mid-century such maps had become common.[100]

One other class of the mapping of physical phenomena had its essential beginnings in the early nineteenth century, namely the mapping of vegetation. Although the plant world has fascinated man for millennia and though the recognition of variations in the distribution of plant types is quite old, the mapping of these phenomena came rather late. Recognition of the effects of temperature on plant distribution, owing to both elevation and latitude, and of the regionalization of plant

*From Single Maps to Atlases*

types, resulted in descriptive studies by Candolle in France, Humboldt and Bonpland in the equatorial regions, Wahlenberg in the Carpathians, and many others in the early years of the 1800s. These usually included only tables and few maps of any kind. Some maps showing limits of various classes of plants appeared early, notably by Carl Ritter in 1805 (FIG. 19).[101] But the pioneering work was that of J. F. Schouw of Denmark, who produced a world atlas in 1823 to accompany his treatise on plant geography.[102]

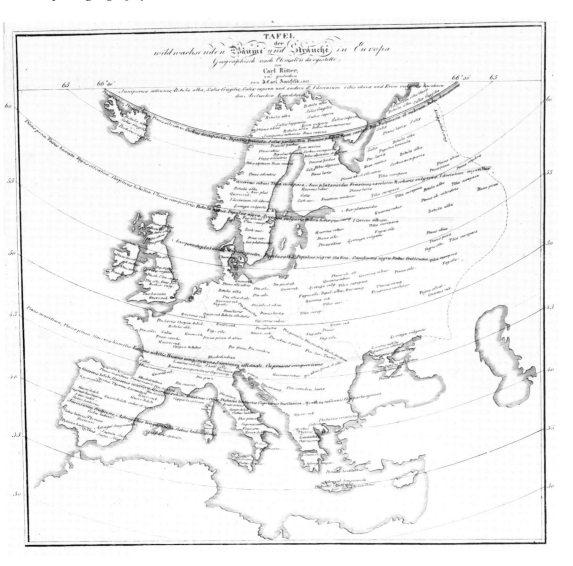

**Figure 19** Ritter's 1805 map of wild trees and shrubs in Europe. Original ca. 1:25,000,000. Copperplate engraving. (Courtesy of the Royal Library, Copenhagen.)

Although other botanical maps appeared thereafter and were nothing unusual by midcentury, thematic zoological maps came much later. Animals, real and imaginary, have been prominent, decorative, space-filling features of maps almost from the beginning, and in spite of the fact that cartographers occasionally departed from zoological truth, their representations appear to have been generally more factual than fanciful.[103] Although maps solely to show zoological distributions were made by E. A. W. Zimmermann in 1777 and by Ritter in the early 1800s, nevertheless, no comprehensive thematic map attempting to show spatial variations of animal life, other than by names or pictorial drawings, appears to have been published until nearly midcentury in Berghaus's *Physikalischer Atlas,* about which more later.

### SURVEY OF NINETEENTH-CENTURY MAPPING OF SOCIAL PHENOMENA

The thematic mapping of the social environment lagged well behind that of the physical. As we have seen, only an occasional such map appeared in the eighteenth century, and it was not until the 1820s that such maps began to appear regularly. Only a few of the milestones can be mentioned here to illustrate the progress. They will be described and illustrated to a greater degree in chapter 5.

One significant observation can be made about this class of thematic maps as compared with those treating physical phenomena, namely, that much of the data was quite different in character, both statistically and geographically. Many natural phenomena are continuous and have precise characteristics of specific locations—for example, temperature, compass declination, and rock character—whereas social and economic data often are derived from areas whose boundaries are unrelated to the data, such as parishes, counties, and *départements.* Furthermore, social and economic phenomena are likely to be discrete, but areally repetitive, such as population. Consequently the thematic mapping of these phenomena called for more innovation and experimentation in symbolization. This may have contributed slightly to its slowness in getting started, but the main reason, as I noted in the last chapter, was that neither the need nor the data came into being until the nineteenth century was well under way. This class of thematic maps will be considered in detail in chapter 6, but a short review is in order in this survey.

Apparently the first such map was made by Baron Charles Dupin in 1826 and published in 1827.[104] It was already relatively sophisticated, showing by shading the number of persons per male child in school in each *département,* in order to show how one part of France differed from another. The same style and type of data were used by Somerhausen a few years later for a map of the Low Countries (FIG. 20). Two years after Dupin, in 1829, A. Balbi, a geographer-statistician, and A. M. Guerry, a lawyer, published in France a set of three maps, one of which dealt

## From Single Maps to Atlases

with the same subject as Dupin's. The other two displayed the distribution of crimes against the person and against property.[105] The maps by Dupin, Balbi, and Guerry were the first to portray "moral statistics" in a long series of such maps to come during the succeeding decades. Maps of moral statistics using a different design for France and the Low Countries, and which were to have wide distribution, were prepared in 1831 by the great Belgian statistician Quetelet to accompany a monograph.[106]

**Figure 20** Somerhausen's post-1829(?) "Carte figurative de l'instruction populaire des Pay-Bas." Original 300 × 350 mm. Lithograph. (Courtesy of the Geografisch Instituut, Rijksuniversiteit Utrecht, Ackersdijck Collection.)

The thematic mapping of population distribution came along at about this same time. To be sure, individual totals simply as numbers had earlier been entered on maps, but more sophisticated representations only began in the late 1820s and early 1830s with such maps as one of population density in Prussia, a small, crude world map of density by Scrope, and a unique dot map of France by Frère de Montizon, all of which apparently went unnoticed until recently.[107] These maps and other maps of population will be treated in detail in chapter 5.

More or less coincident with the introduction of thematic maps of moral statistics was another group of maps concerned with man's ills, specifically his problems of health. The great cholera epidemics which ravaged western Europe gave rise to a great many maps of all varieties having to do with the distribution of that disease and with those factors thought to be involved.[108] They were of all scales and ranged from parts of cities and whole cities to countries. They often accompanied medical reports and surveys and became common by midcentury. Some were quite innovative, and except for those by August Petermann, the great German cartographer-geographer, most were conceived by medical practitioners.

The last of the categories of thematic maps to become popular by the mid-nineteenth century constituted the economic maps displaying the production and movement of goods. Apparently the first flow map was printed in 1837, but it was not until the 1840s and 1850s that such maps were being produced regularly.[109] Thereafter, they also became common, and because they came relatively late they could take advantage of the greater use of printed color, a technique which had evolved by then.

### Thematic Atlases

The great number of thematic maps that began to appear in the nineteenth century, like their predecessors, were often appended to reports and monographs, appeared as illustrations in books or, very occasionally, were published separately. This was unlike many general maps, which were either issued separately or had been systematically produced in a single format and bound as sets, called atlases since the sixteenth century. To be sure, Guettard and Monnet had taken advantage of the new "Carte de Cassini" of France and had produced a mineralogical atlas in the latter half of the eighteenth century, which probably qualifies as the first thematic atlas even though its primary function was simply to show where various minerals were to be found.[110] The atlas included forty-five sheets covering northern France. A portion of one is shown in FIGURE 36 in chapter 4.

The first printed atlas devoted to the interrelationships among geographical phenomena, and thus clearly thematic, was the small 1806 *Sechs Karten von Europa mit erklärendem Texte...* by Ritter. It was prepared for instructional use, and as time went on similar teaching aids began to appear, almost always as a series of general maps to which one or two thematic maps would be added. This practice continued to develop and by midcentury had matured a great deal. For example, the National Society for the Education of the Poor in England and Wales published in 1852 an atlas containing twelve maps, of which at least half are clearly thematic.[111] These include mineral production, temperature, precipitation, population, and chief occupations. Ten years later, Maurice Block published a purely thematic atlas for general distribution to accompany a book on the comparative strength or power of the several countries of Europe.[112] It contained twelve maps, ranging in subject from population density and public debt to customs duties and dominant religions. Other than coastlines, the only physical phenomena displayed were navigable waterways.

In the first part of the nineteenth century a few "manuscript" atlases were prepared in which thematic data were drawn on sets of printed base maps. Examples are Schouw's *Pflanzengeographischer Atlas...* (Schouw 1823b) and the Prussian *Administrativ-Statistischer Atlas...* (Prussia 1828). Although such thematic atlases and the thematic component of school atlases demonstrate that mapping all kinds of subject matter was acceptable, they seem but minor achievements when compared with the *Physikalischer Atlas* of Heinrich Berghaus and its later English version, A. K. Johnston's *The Physical Atlas*. The former can be thought of as the culmination of all the scientific-cartographic innovations in ways of looking at the physical world and the techniques for its portrayal that had been evolving since the times of Kircher and Halley.

The *Physikalischer Atlas* had its origin in Alexander von Humboldt's plan to write a complete physical geography *(Kosmos)* and his feeling that it would need a thorough atlas of physical phenomena as an accompaniment. Humboldt wanted Berghaus, one of Europe's leading cartographers, to undertake the atlas with the Cotta publishing house, with which both Humboldt and Berghaus had connections.[113] To make a long story short, Berghaus in Berlin and Cotta in Stuttgart got into a controversy in 1827 and in the end parted company for good.[114] Berghaus then began to negotiate with Justus Perthes and Adolf Stieler in Gotha regarding an atlas, possibly to be an addition to Stieler's *Hand Atlas*, but that idea did not materialize. Work did get under way, but it was interrupted for several reasons until 1836. In 1839 Berghaus moved from Berlin to Potsdam and opened his *Geographische Kunstschule*, and work on the atlas proceeded both at Potsdam and at Perthes's geographical institute in Gotha. The *Physikalischer Atlas* ultimately appeared in two volumes, the first in 1845 and the second in 1848. The individual maps are dated variously, beginning in 1838.

The contents of the *Physikalischer Atlas* are divided into eight sections as follows:

| *First volume (1845)* | *Pages of maps* |
|---|---|
| 1. Meteorology and climatography | 13 |
| 2. Hydrology and hydrography | 16 |
| 3. Geology | 15 |
| 4. Earth magnetism | 5 |
| 5. Botanical geography | 6 |
| *Second volume (1848)* | |
| 6. Zoological geography | 12 |
| 7. Anthropography | 4 |
| 8. Ethnography | 19 |

Berghaus included extensive explanations and notes to the maps covering such matters as sources, theory, and history of the subjects, in which he was aided from time to time by Humboldt.

The atlas sheets began to appear in 1838 and were issued as a supplement to Stieler's *Schul Atlas* in 1839. In 1841 a connection was established with the publishing house of J. Johnstone, W. and A. K. Johnston in Edinburgh through the German ethnographer, Kombst, living in Edinburgh, who was compiling an ethnographic map for Berghaus. Although scientific physical geography was not nearly so advanced at that time in Britain as in Germany, W. and A. K. Johnston were anxious to produce an English physical atlas. In 1842 A. K. Johnston entered into an agreement with Berghaus to publish a few maps on trial. These were incorporated in Johnston's *The National Atlas* in 1843 and followed in 1845 by Johnston's *The Physical Atlas*. To assist in the preparation of the large English version of the *Physikalischer Atlas*, two of Berghaus's student apprentices who had finished their training in Berghaus's school went to Edinburgh, H. Lange in 1844 and August Petermann in 1845.

The revised, updated, and reengraved maps for A. K. Johnston's large *The Physical Atlas*, "Based on the Physikalischer Atlas of Professor H. Berghaus..." and dedicated to Alexander von Humboldt, appeared in 1848.

Berghaus's *Physikalischer Atlas* was a monumental achievement, bringing together an enormous amount of information about the physical geography of the earth and encompassing many more subjects than had been treated in the numerous smaller atlases that had been produced since the 1820s. Nevertheless, as Berghaus himself said, his atlas was a "collection" of maps of differing formats, employing a variety of techniques and made over a ten-year period. As Engelmann points out, the balance required in an atlas was lacking, and it was A. K. Johnston's contribution to bring this wealth of material together, distill it, revise it, and execute all the maps in a unified manner in the first truly comprehensive thematic atlas.[115]

*From Single Maps to Atlases*

It is fitting to end this survey of the growth of thematic cartography with a glimpse of the life and influence of Alexander von Humboldt. His place in the history of thematic cartography is curious.[116] On the one hand this greatest universal naturalist of the nineteenth century made few thematic maps, more often creating diagrams than maps. On the other hand, he made a monumental contribution by transforming the Halleyan line into an isotherm from which issued scores of cartographic offspring; his cross sections and geological mapping, including symbols for strike and dip, were innovative, and he was the main force behind the great *Physikalischer Atlas* of Berghaus and its English version, *The Physical Atlas* of A. K. Johnston. Although Humboldt himself published relatively few thematic maps, he probably provided the data for and inspired more of them than any one individual ever has done.

It would be inappropriate in this book to devote much space to the life and travels of Humboldt, which have been extensively treated by numerous students of his career and his importance in the development of geography and science.[117] It is important to realize that he spanned the entire period when thematic cartography was maturing. He was born in 1769 and began his academic studies in 1787. He had a brief career in public service as an assessor in the Prussian Department of Mines, and in 1794 he published his first major work dealing with subterranean plants in mines. He did not want a career in mining, and from 1799 to 1804 he and Bonpland, a botanist, traveled in the New World, from the tropics of Latin America to the United States, after which he settled in Paris, the center of the scientific world. He was required to go to Berlin, but he returned to Paris in 1808, where he remained (on a pension from the Prussian government) until 1827. During those nearly twenty years he moved in the highest scientific circles and assumed his position as the leading naturalist of his time. After returning to Berlin in 1829 he took a long journey across Russia to the Chinese border at the invitation and expense of the czar, covering more than ten thousand miles—at the age of sixty. After his return he was engaged in writing of various kinds for the remainder of his life. He began to write *Kosmos* in 1833, when aged sixty-five, and was working on the last volume in 1859 when he died at age ninety.

Humboldt concentrated on natural history, and in that broad area his career and the development of thematic cartography are coincident. It may not be an overstatement to say they are nearly synonymous.

*Four*

# MAPS OF THE PHYSICAL WORLD

As I noted previously, the physical world claimed most of the attention in the early period of the era of thematic mapping beginning about the mid-seventeenth century. This was a time of overseas expansion, with extensive maritime exploration, and the scientific arts of navigation and mapmaking were intimately involved. The practical needs of navigation, such as determining a position, following a course, or making allowance for currents, depended not only upon the accumulation of widespread observations but just as much on the development of theory which would make possible prediction and the testing of hypotheses. For both navigation and mapmaking, the major areas of scientific interest were astronomy, geomagnetism, currents in the atmosphere and oceans, and the shape of the earth.

As time went on the character of the rocks and the structures underlying the surface became elements of great interest because they were primary factors in mining, drainage, and canal building as well as for their significance in deflecting the plumb line used in mapping. As the interrelationships among the distributions of vegetation, animal life, temperature, precipitation, and man began to appear, such components as the elevation of the land surface and the occurrence of mountain systems also increased in importance. Each of these led to thematic mapping in one way or another.

The growth of an immature science and technology is rarely smooth and direct. In historical perspective it often appears to be erratic as interests shift and new techniques open the way to experimentation and to developments in adjacent fields.[118] This seems to have been particularly so in the earth sciences, partly because the subject matter is so broad and partly because the first individuals involved were remarkably versatile, often being unusually competent in several fields. Halley and Humboldt are outstanding examples.

*Maps of the Physical World*

To characterize in condensed fashion the development of thematic mapping of the physical world from the mid-seventeenth century to the last half of the nineteenth, by which time most of the cartographic developments had taken place, involves several problems and risks. Since only a few of the maps can be illustrated, those chosen ought to represent fairly the degree of activity in each category; not an easy task, and made more difficult by the practical and technical problems of availability and reproduction. Probably more important from the cartographic point of view is maintaining the focus on the maps themselves and their objectives, and the techniques for attaining them, rather than upon the often fascinating development of knowledge about the subject matter. Third, one must keep in mind that, ideally, the thematic map has as its purpose the display of the overall structure or meaningful character of a distribution rather than simply showing where some things are. The distinction is sometimes difficult, since the range from the thematic to the general map is a continuum and most maps have qualities of both.[119] I will begin this survey of the development of the thematic mapping of the physical world with a look at the portrayal of atmospheric phenomena.

THE ATMOSPHERE

In chapter 2 I pointed out that Halley published what has been called the first meteorological chart in the *Philosophical Transactions* for 1686.[120] It is shown in its entirety in FIGURE 21 at a reduced scale.[121] Halley obtained the data both from his own observations while spending nearly a year on the island of Saint Helena in the South Atlantic working on a star chart of the southern heavens, and from gleaning the observations of ships' masters and others. He understood the value and objective of such mapping by observing that the phenomenon "is better expressed in the Mapp hereto annexed, than it can well be in words." As Thrower points out, crude as the map may appear, it has qualities that fit the aims of thematic mapping better than others made later that superficially appear visually more attractive.[122] The chart employs Mercator's projection, the conformal property of which makes it possible to portray the winds with their correct bearings anywhere.

Halley's map was occasionally reproduced during the next forty years, and the wind system was displayed on many English and Dutch nautical charts of the first third of the eighteenth century.[123] A decade later William Dampier published two maps, one of the Atlantic Ocean and Indian Ocean and one of the Pacific, showing the trade and monsoon winds.[124] Apparently, such other maps of the winds as were made during the eighteenth and early part of the nineteenth century were on nautical charts and are essentially nonthematic, and it was not until the

1840s and the appearance of Berghaus's *Physikalischer Atlas* and Johnston's *The Physical Atlas* that the wind systems were treated in more than a cursory thematic fashion.[125]

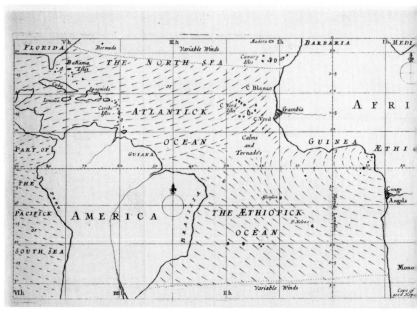

**Figure 21** Halley's 1688 map of the trade winds. Original 489 × 146 mm. Copperplate engraving. The map has neither title nor legend. (Courtesy of the British Library.)

Since the beginning of the seventeenth century thermometry had been developing, and by the mid-eighteenth century the Fahrenheit, Réaumur, and Celsius (centigrade) scales had all been devised. By the last quarter of the eighteenth century instrumentation had become good enough that comparable measurements could be obtained at different places by different observers. The decrease of temperature with increasing altitude and latitude was known, of course, and the relationship among altitudinal and latitudinal variations in temperature, the snowline, and vegetation was appreciated. Carl Ritter, for example, included such mountain profiles as well as a map of Europe in his *Sechs Karten von Europa*, published in 1806.[126] Toward the latter part of the 1700s observations of temperature and other meteorological phenomena were being kept at various places and for varying periods, occasionally appearing on maps. Simply putting numbers on maps, such as temperatures at various places, is unsatisfactory except for reference purposes, and the thematic mapping of temperatures and several other atmospheric phenomena required the development of a new cartographic technique. Alexander von Humboldt accomplished this in

1817 by adapting the Halleyan concept of "curve lines" of equal compass declinations to show the distribution of temperatures by means of isotherms.

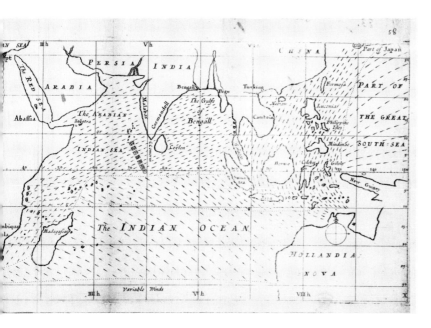

Humboldt's small, diagrammatic map of isothermal lines of average annual temperatures, FIGURE 22, attempts to portray a basic characteristic of zonal temperature distribution, namely, that the rate of decrease with increasing latitude differs on the eastern and western sides of continents. It had great impact in spite of the fact that it did not appear with the *mémoire* it illustrated or even in the same journal.[127] A translation of Humboldt's *mémoire* appeared, without the map, in the *Edinburgh Philosophical Journal* in 1820 and 1821, and even as far away as the New World named isotherms were used on a map in a school atlas published in Connecticut in 1826.[128] Humboldt's map was reproduced in Germany in 1827, and the technique was adopted by Kämtz in 1832 and Mahlmann in 1836 for the preparation of more detailed maps of annual isotherms for which data had by that time become sufficient.[129] The first map in Berghaus's *Physikalischer Atlas* is entitled "Alexander von Humboldt's System der Isotherm-Kurven in Merkator's Projection" (FIG. 23). The map was prepared in 1838, and the isotherms are highly generalized. Only one set of lines is used to show the distribution of both summer and winter temperatures, this being done by placing the appropriate values for winter on the poleward side of the lines and those for summer on the equatorward.

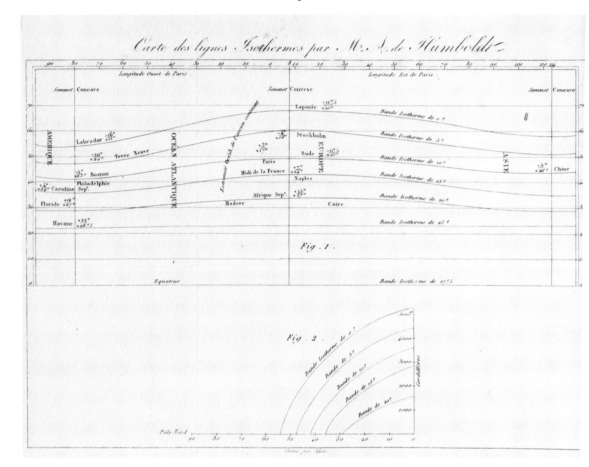

**Figure 22** A. von Humboldt's 1817 chart of isothermal lines. Original 219 × 95 mm. Copperplate engraving. (Courtesy of the British Library.)

The combination of Humboldt's fame as the foremost natural scientist of his time and the utility of the isoline as a device with which to display quantitative geographical phenomena led to the widespread use of the technique in thematic mapping. Humboldt himself in his original paper had suggested lines of equal winter and summer average temperatures, which he termed "isotheres" and "isochimenes." By giving the lines specific names rather than simply referring to them generically, as Halley had done with his "curve lines," Humboldt unwittingly initiated a proliferation of jargon, the ultimate extent of which probably would have pained him.[130]

Maps of the Physical World

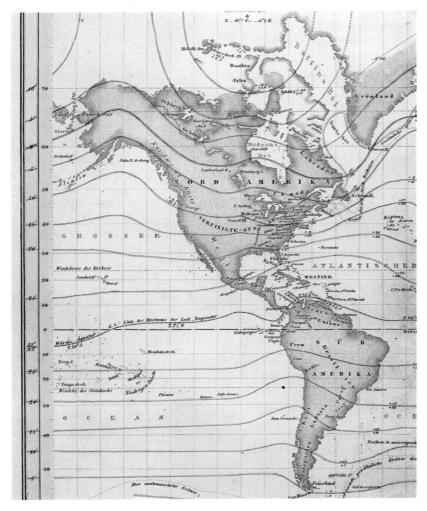

**Figure 23** Part of the 1838 world map of isothermal lines in Berghaus's *Physikalischer Atlas,* first Abteilung, Meteorology, no. 1. Reduced 35 percent. Copperplate engraving, hand colored. (Courtesy of the British Library.)

One of the topics to which the line of equal value soon found application was atmospheric pressure. Since the mid-seventeenth century when Toricelli devised the barometer, observations of atmospheric pressure had been made for various reasons, mostly scientific, but variations over the surface had not been mapped. In 1820 H. W. Brandes of Breslau, investigating the relation between winds and air pressure, made a synoptic map of conditions on 6 March 1783, when an unusually strong, disastrous storm had afflicted western and central Europe. He plotted the deviations from normal pressure and joined equal deviations with lines, the data being available from the records of the Societas Meteorologica Palatina at Mannheim.[131] That institution, founded in 1780, had members in many different countries, encouraged them to keep records, established rules for doing so, and issued a journal. The relation between wind direction and barometric pressure was readily apparent. In 1827 Kämtz used "isobarometric lines," lines of equal barometric variation,[132] and later he included a map of isobarometric lines of the earth in the second volume (1832) of his three-volume *Lehrbuch der Meteorologie*.[133] Isobars, lines of equal average pressure, conceived as parallels of latitude rather than as irregular lines, were first

**Figure 24** Olsen's 1839 "Carte hyetographique générale" in *Atlas pour le tableau du climat de l'Italie*. Full size. Lithograph, crayon shading. (Courtesy of the Royal Library, Copenhagen.)

*Maps of the Physical World*

used by Berghaus in the *Physikalischer Atlas* on a map dated 1839. Mapping irregular lines of barometric deviation was done by Loomis in the United States in 1846, but it was not until 1864 that the modern concept of the isobar as an irregular line joining points of equal air pressure, not paralleling parallels, was mapped by Renou.[134]

The earliest map of precipitation now known is a shaded map of parts of Europe and Africa dated 1839, in an atlas prepared by the Danish cartographer O. N. Olsen to accompany a study of the climate of Italy by the botanist Schouw.[135] The atlas contains five plates, one being a hachured map of Italy, two being temperature maps, and two being "hyetographic" or precipitation maps. One temperature and one precipitation map are of Italy, and they show only numbers, being therefore more nearly general maps for reference. The other two cover parts of Europe and Africa and are thematic. The temperature map shows annual isotherms with a 2°C interval. The precipitation map is shown in FIGURE 24 and is the more interesting cartographically, since it uses shading to show seasonal concentration in four east-west *bandes* (continuous rain, winter rain, without rain, summer rain), with additional shading in appropriate places to show the orographic effect. The map is accompanied by a clear and concise explanation.

Berghaus's *Physikalischer Atlas* includes a world hyetographic map, dated 1841, on which shading was employed in a more realistic manner than on Olsen's map. A portion is shown in FIGURE 25, and this appears to be the earliest world map of precipitation. In addition to the variable shading (the darker, the greater precipitation), the map includes colored lines to show limits of various zones of seasonal concentration and snowfall, and it shows wind arrows in the Indian monsoon area. In the commentary Berghaus points out that there is insufficient data to make an isoline map of annual precipitation of the world.

The *Physikalischer Atlas* does contain a map of Europe, dated 1841, with "curves" of equal annual amounts of precipitation, a portion of which is shown in FIGURE 26. It portrays considerable additional information, such as "curves" of summer rain consisting of dashed lines with dots interspersed, each dot representing 5 percent of the annual total. It shows the directions of "rain winds" and, for selected stations, the number of rainy days, the amount of precipitation on one day, and the number of days with snow. Johnston's *The Physical Atlas* includes a "Rain Map of the World" similar to Berghaus's, as well as one of Europe, and these set the pattern for the future.

Petermann, the German cartographer-geographer who had worked on maps for both the *Physikalischer Atlas* and *The Physical Atlas*, produced his own *Atlas of Physical Geography* ... in 1850.[136] It contains fifteen copperplate maps and 106 pages of text on physical geography by Thomas Milner illustrated by 130 "Vignettes on Wood," excellent wood

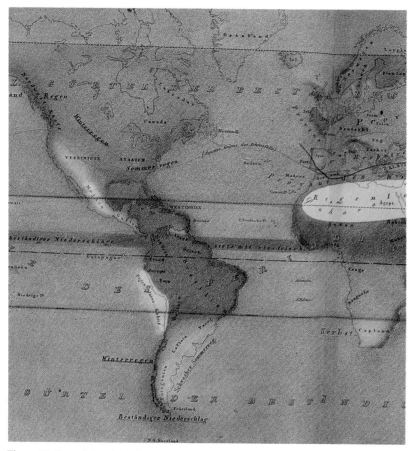

**Figure 25** Part of the 1841 "Hyetographische Karte der Erde" in Berghaus's *Physikalischer Atlas*, first Abteilung, Meteorology, no. 9. Reduced 37 percent. Copperplate engraving, aquatint, hand colored. (Courtesy of the British Library.)

engravings of sections, views, diagrams, structures, and so forth. The maps are noted as being constructed by Petermann, but they were "Engraved by J. Dower of Pentonville, London," a most competent craftsman. FIGURE 27 is a portion of the "Hyetographic Map Showing the Distribution of Rain over the Globe," and Petermann's description is instructive.

> This map represents, by different tints of shading, the comparative amount of rain that falls in various localities of the globe: the deepest tints denoting the greatest quantity, and the blank spaces the rainless districts. The enormous diminution of the average

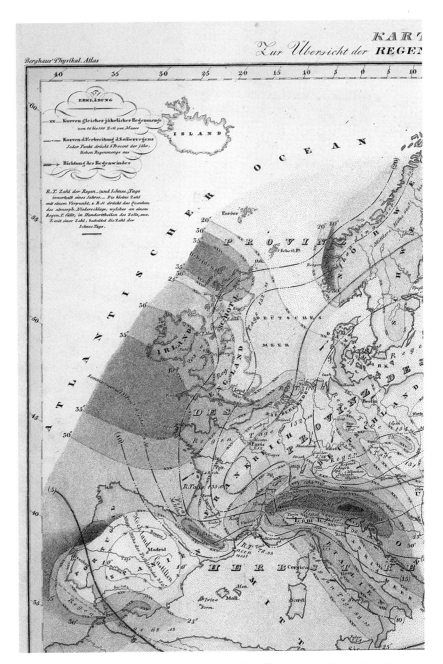

**Figure 26** Part of the 1841 "Karte von Eüropa, Zur Übersicht der Regen-Verhältnisse in diesem Erdtheile" in Berghaus's *Physikalischer Atlas*, first Abteilung, Meteorology, no. 10. Reduced 33 percent. Copperplate engraving, hand colored. (Courtesy of the British Library.)

quantity of rain from the equator to the poles will be observable at a single glance....

The shading, though affording an extremely striking and conspicuous general view of the amount of rain, is quite insufficient to represent accurately the various tints ranging between 1 and 300; therefore *figures* have been inserted in the Map, expressing definitely the average quantity of rain observed in various parts of the globe.[137]

One curious attempt at another form of representation, which suggests the degree to which symbolic experimentation was still very much the order of the day in thematic cartography, is the "Hyetographic Map of the British Isles and the Neighboring Countries...," dated 1851, included in the "National Society" atlas.[138] The part showing the British Isles is shown in FIGURE 28, and in the preface the map is described as follows:

the quantity of rain which falls in particular spots is indicated by the *size* of the blue circles, and not by their number. Their number shows only the places at which accurate observations are recorded.

Designed by Samuel Clark, the map was drawn by Petermann. The technique did not survive.

The variety of atmospheric phenomena that can appropriately be displayed on maps thematically is very large indeed, and no attempt is here made to identify the first instance when each was mapped, since the cartographic methodology was not particularly new.[139] By the 1850s the isoline and graduated shading were essentially standard symbolism. Thereafter changes were largely a result of technical and conceptional sophistication, but in general the cartographic portrayals were not innovative.[140]

The thematic mapping of individual meteorological phenomena was quickly followed by attention to a much more difficult cartographic problem, the weather map, as we call it today. The first of these appear to be the set of storm maps by E. Loomis published in 1846.[141] On his maps, which were presented to the American Philosophical Society in 1843, in addition to wind direction and force, Loomis showed cloudiness, precipitation—both rain and snow—and the distribution of air temperature and pressure. He justifiably stated:

Nearly every circumstance essential to a correct understanding of the phenomenon of the storm is thus presented to the eye at a single glance.

Weather maps are complex thematic maps which attempt to integrate

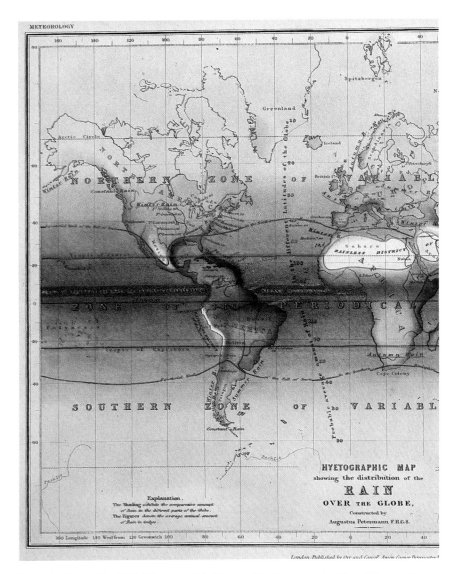

**Figure 27** Part of the 1850 "Hyetographic Map . . ." in Petermann's *Atlas of Physical Geography*. Reduced 33 percent. Copperplate engraving, aquatint, hand colored. (Courtesy of the British Library.)

graphically the structures of several phenomena, such as winds, precipitation, temperature, and so on. Hellmann reports that weather maps began to be issued first, temporarily, at the International Exhibition in London in 1851, and by the early 1860s had become standard in both England and France.[142]

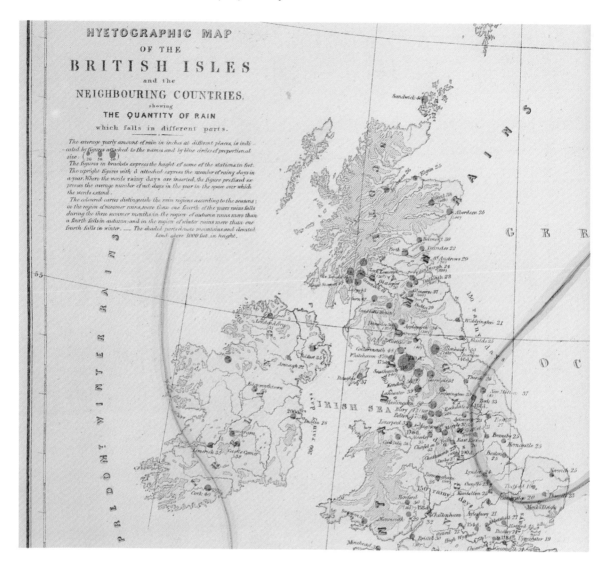

**Figure 28** Part of the 1851 "Hyetographic Map..." in the National Society's atlas *Maps Illustrative of the Physical, Political and Historical Geography of the British Empire*. Reduced 10 percent. Lithograph, blue circles printed color, other color by hand. (Courtesy of the British Library.)

## The Oceans

The early thematic treatment of currents in the ocean fared not much better than did the winds of the atmosphere. The earliest maps of ocean currents are by Kircher, who included one world map and six other maps of parts of the Atlantic Ocean in his *Mundus Subterraneus*.[143] The world map is shown in FIGURE 29. Kircher's maps (and Happel's, who followed him, see FIG. 11) show no sense of direction, such as by arrows, so that one would need to read the text to learn the nature of the circulations. Eckert makes reference to arrows on an anonymous chart of 1739,[144] but the systematic representation of ocean currents seems not to have been undertaken again until the last quarter of the eighteenth century.

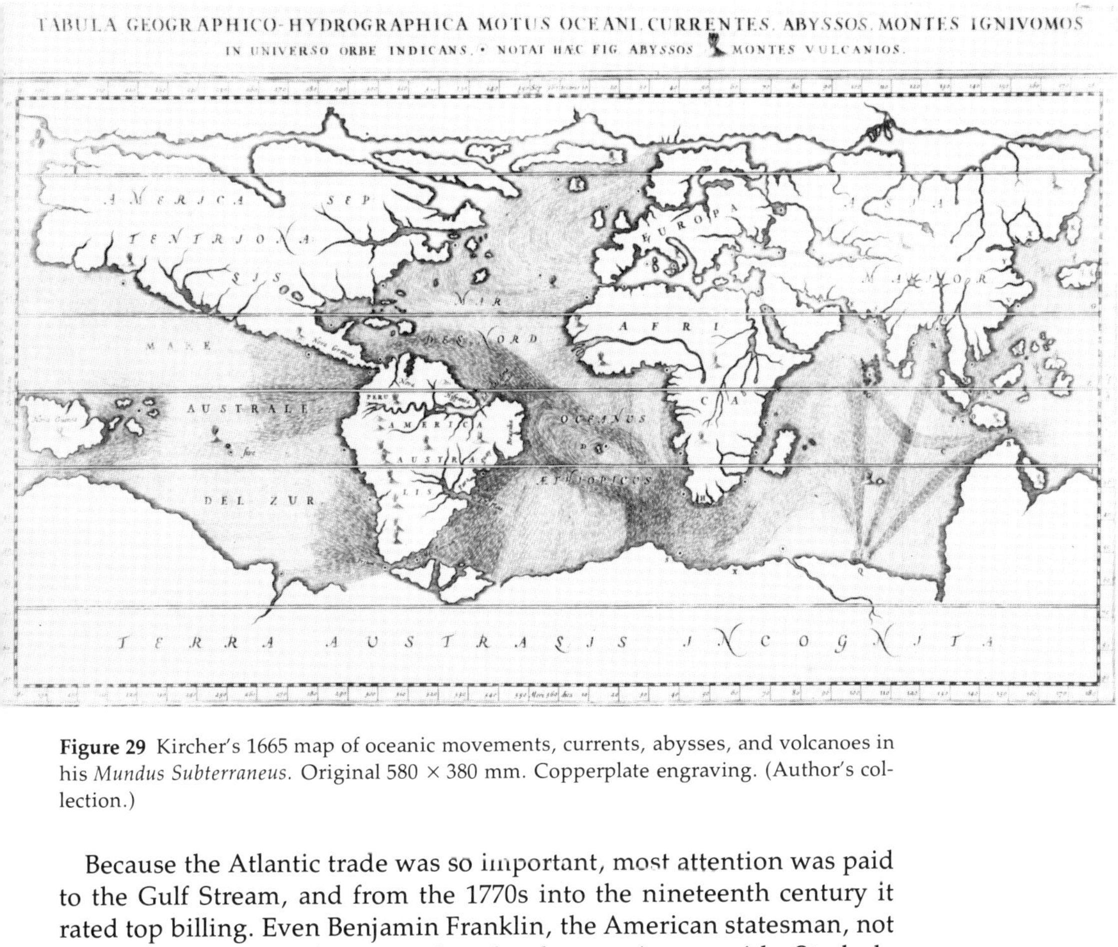

**Figure 29** Kircher's 1665 map of oceanic movements, currents, abysses, and volcanoes in his *Mundus Subterraneus*. Original 580 × 380 mm. Copperplate engraving. (Author's collection.)

Because the Atlantic trade was so important, most attention was paid to the Gulf Stream, and from the 1770s into the nineteenth century it rated top billing. Even Benjamin Franklin, the American statesman, not known as a cartographer, joined such others as Arrowsmith, Steel, de Brahm, and Delaméthrie in charting that current.[145] Franklin's contribution began in 1768, as reported by De Vorsey, but the extant maps ascribed to him date from 1785. FIGURE 30 is the chart published in 1786 attributed to Franklin.[146]

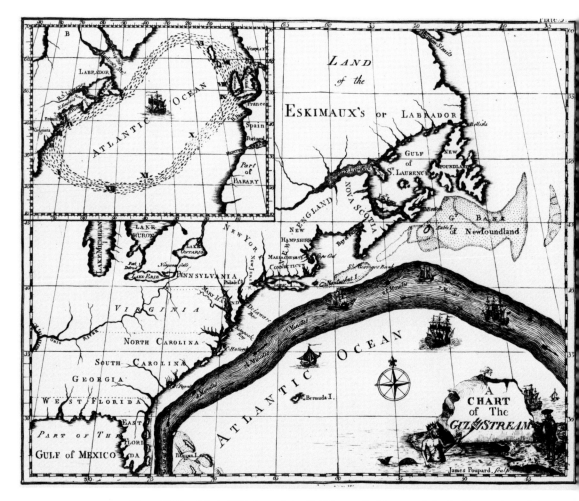

**Figure 30** Franklin's 1786 chart of the Gulf Stream in his memoir in the *Transactions of the American Philosophical Society*. (The smaller chart in the upper left illustrates Gilpin's paper on the seasonal migration of herring.) (Courtesy of the University of Wisconsin Memorial Library.)

Perhaps the reason currents in the open ocean were rather slow in being mapped was the difficulty of their observation, as well as the lack of widespread travel, except in limited areas such as the Atlantic. Wind direction and velocity are easy to observe, but when one is in a craft being moved by the wind any relative movement of the water is impossible to observe without a fixed reference point. Currents near land could be observed, and mapping of the tidal sets was done relatively early, for example, by Halley. Thematic maps of ocean currents for whole oceans and the world began to appear by the time of the

*Physikalischer Atlas* and *The Physical Atlas,* while in Johnston's 1852 *Atlas of Physical Geography...* the initial map of the world showed ocean currents.[147] Maps of several kinds of physical phenomena were often combined in one map as, for example, in Petermann's "Hydrographical Map of the World Showing the Currents, Temperatures...of the Ocean" (FIG. 31).[148]

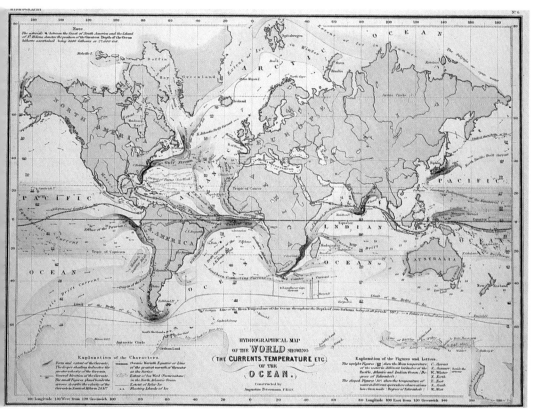

**Figure 31** The 1850 "Hydrographical Map of the World..." in Petermann's *Atlas of Physical Geography.* Original 280 × 205 mm. Copperplate engraving, hand colored. (Courtesy of the British Library.)

GEOMAGNETISM

Geomagnetism became a subject for thematic mapping when it began to be realized that the departure of the compass needle from a true north-south line (declination or variation) might be systematic. It was readily apparent that if a regular pattern of distribution did exist, it would be extremely helpful to mariners to know about it, for two reasons. First, they could correct their bearings, and, second, if it varied with longitude they might be able to reckon their east-west position, something they could not do easily in any way.

The mariner's compass may have been invented in China perhaps nearly nine hundred years ago, and apparently the Chinese observed and recorded declinations quite early.[149] The compass began to be used in the Western world about the beginning of the twelfth century, but seemingly the knowledge of declination did not accompany it. When European ships began moving over the open ocean after the fifteenth century the fact of declination was apparent—and bothersome. Its mapping became important, for a mariner's life did indeed depend upon knowing where one was on the sea.

The known early attempts at mapping declination include the manuscript charts by Alonso de Santa Cruz of Spain in 1536, by Luis Teixeira of Portugal about 1585, by the Italian Jesuit Christoforo Borri in the early seventeenth century, and by the Jesuit Martinus Martini on a world map ca. 1630–40.[150] There was much study of the problem of the observing of and the correcting for declination during the fifteenth, sixteenth, and seventeenth centuries.[151] Kircher, the Jesuit scholar at Rome, wrote in 1654 that he had enough data to make a proper isoline map of declination but that cost and time did not allow it to be done to be printed for his book.[152] The very considerable honor of being the first to produce a printed isoline map of declination belongs to Edmond Halley, who made a voyage to the south Atlantic to gather data in 1698–1700, then produced a thematic map of declination of the Atlantic Ocean in 1701 (FIG. 32). The following year he produced a much larger, similar map of the world not including the Pacific Ocean, for which he said not enough data were available.[153]

As stated by Halley in an explanation to accompany the maps, the purposes of his chart were to allow the mariner to correct his course especially "in continued cloudy weather" and to estimate the longitude in areas where the "Curves run nearly North and South, and are thick together." Halley referred to the isolines of declination by stating in the explanation:

> What is here properly *New*, is the *Curve Lines* drawn over the several Seas, to shew the degrees of the *Variation* of the *Magnetical Needle*, or *Sea-Compass*.

The term "curve" or "curve line" was nothing new, simply being the general designation, by mathematicians of the time, for a line along which some characteristic remained constant.[154] What was "properly new" was the positions of the lines, since, as Halley stated, they were based upon his own original observations in the Atlantic and upon journals of voyages in "the *Indian* Seas" corrected to 1700. Halley did not invent the concept of isolines on a map, but he certainly made the concept generally known through the publication of his maps, which

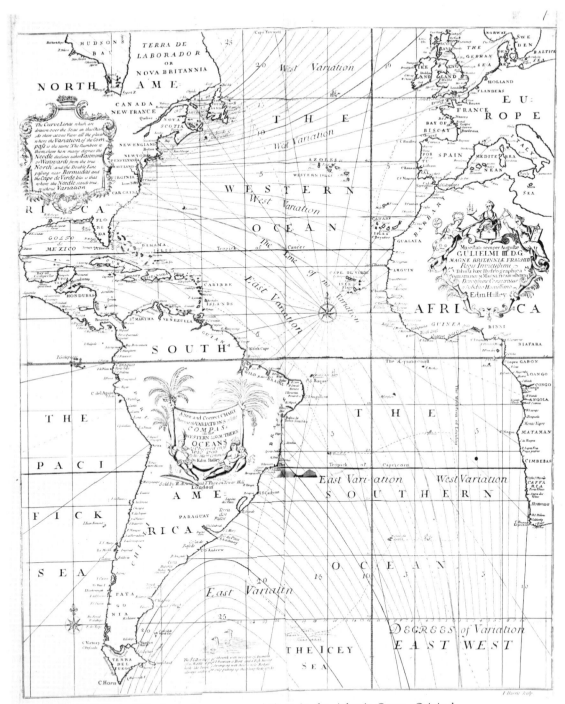

**Figure 32** Halley's 1701 chart of compass variations in the Atlantic Ocean. Original 570 × 482 mm. Copperplate engraving. (Courtesy of the Library of Congress.)

were widely disseminated during the 1700s.[155] The curve lines were called Halleyan lines by many.[156]

During the century and a half after Halley's charts were made available, the study of geomagnetism expanded steadily, and a variety of thematic, isoline maps were produced. Whiston made a map of inclination of the magnetic needle in 1721; J. C. Wilcke of Stockholm made a world map of inclination in 1768; Humboldt mapped (diagrammed) isodynamic zones—that is, variations in field strength or the intensity of gravity—in 1805; C. Hansteen mapped isodynamic lines in 1825 and 1826; and in the following decade L. I. Duperrey of France (1833) and E. Sabine of England (1837) published isodynamic maps of the entire earth. Duperrey's maps were incuded in Berghaus's *Physikalischer Atlas*, dated as 1837 and 1839.[157]

### Geology

Geognostic (geological) mapping had its beginnings in the first half of the eighteenth century. Rocks, and especially minerals, have fascinated man since before Greek and Roman days, but those earthly phenomena had no operational significance, as geomagnetism did for navigation. Consequently there was no strong urge to map their distributions except as the individual natural historian became fascinated by the arrangements of phenomena and mapped them to see their relationships.

The oldest purely geological society, that of London, was founded in 1807, but learned societies such as the Royal Society and the Académie had been receiving reports on geognostical subjects since before the first half of the eighteenth century. Again, in Britain the first official geological survey was founded in 1835, though France had earlier delegated to two geologists the preparation of a geological map of all of France. Most other official surveys were founded considerably later.

Geological maps generally combine characteristics of both the general and the thematic, and it is often difficult to place them in one category or the other. For example, what is usually described (by geologists) as "the earliest geological map," that of Christopher Packe, referred to in chapter 3, is at a scale of about 1:42,000 (ca. 1.5 inches to the mile), which is large scale for a thematic map. Often such early geological maps were simply base maps to which had been added some symbols to show the locations of various rock types or minerals, and such maps made no attempt to portray the character of a distribution. Maps showing various formations, and especially those relating them to the landforms, would be classed as thematic, except when they are of such a large scale as to be useful primarily for reference purposes.

The earliest thematic geological maps, other than those of Packe, were

## Maps of the Physical World

made by Philippe Buache, some for himself and some to illustrate the geological work of Guettard which was briefly noted in chapter 3.[158] Buache was born in Paris in 1700 and first distinguished himself in the art of drawing and won a prize for architecture. The geographer-cartographer Guillaume Delisle took him in, and Buache thereafter devoted himself to geography and cartography.[159] In 1721 he was appointed an official of the Dépôt de la Marine, where he spent seventeen years putting in order the store of maps, plans, and marine log books. His salary was not large, but even though the astronomer J.-N. Delisle, Guillaume's brother, tried to entice him to Russia, he elected to remain in France. G. Delisle, *premier géographe du roi*, died in 1726 and in 1729 Bauche was granted that title. A place in the Académie for a geographer was created for him, and he became a member in 1730.

In 1746 Buache prepared two maps to accompany an article by Guettard, and these should rank as the first truly thematic geological maps.[160] Although dated 1746, the maps did not appear until 1751 when the *Mémoires* of the Académie for 1746 appeared. The first map is shown in FIGURE 33. It includes forty-nine symbols for minerals and rocks, but, more important, it shows, by shading and lettering, the division of the area into three zones or *bandes*—an outer one called *schiteuse ou metallique* surrounding a marl-clay zone called *marneuse*, which in turn encompasses a sandy zone, a *bande sablonneuse*. The second map is simply a larger-scale center section of the first. As noted in chapter 3 Buache prepared three similar maps to accompany papers by Guettard—one of Switzerland, one of the Middle East centered on Egypt, and one of eastern North America.[161]

Two other thematic maps by Buache are also geological and accompanied a paper presented to the Académie in November 1752.[162] The first map, shown in FIGURE 34, illustrates Buache's suggestion that the earth's major mountain ranges might be connected by extensions beneath the oceans. Although he emphasized the lack of data to support the idea, his proposal found little acceptance.[163]

The other map is far more significant in the history of cartography. It also appeared with his essay of 1752 and is shown in FIGURE 35. In that map he was portraying the connection between France and England, to which he appended at the top a profile, with depths, along the Channel and into the North Sea. The important aspect of the map, cartographically, is that Buache used isolines of depth (since called isobaths), in the open ocean and points out, on the map, that he had presented this in manuscript to the Académie in 1737. In the text he also points out that he had used this technique in 1736 on a map, *l'Océan vers l'equator*, which had been kept at the Dépôt de la Marine. He also claimed to be the first to have made such use of soundings, but that is doubtful. His claim and objectives were stated as follows:

The use I have made of the soundings, and which no one has employed before me to convey the depths of the sea, seems to me very appropriate to make known in a sensible manner the gradient or slope of the coasts, and ... which shows us by degrees ... the bottoms of the basins of the sea.[164]

**Figure 33** Buache's 1746 "Carte Minéralogique sur la nature du terrain ... de l'Europe." Original 235 × 320 mm. Copperplate engraving. (Courtesy of the University of Wisconsin Memorial Library.)

Maps of the Physical World

**Figure 34** Buache's 1752 "Planisphère physique..." showing "the great mountain ranges which span the globe." Original 324 × 312 mm. Copperplate engraving. (Courtesy of the University of Wisconsin Memorial Library.)

Even though Buache may not have been the first to employ the technique in the open ocean (isobaths had been used previously in rivers and estuaries) the appearance of his map in the *Mémoires* of the Académie certainly helped greatly to spread an understanding of the technique. A simplified rendition of the map was made by Buache to illustrate a *mémoire* by Desmarest, one of France's notable eighteenth-century geologists, on the subject of England and France being part of a single continent.[165]

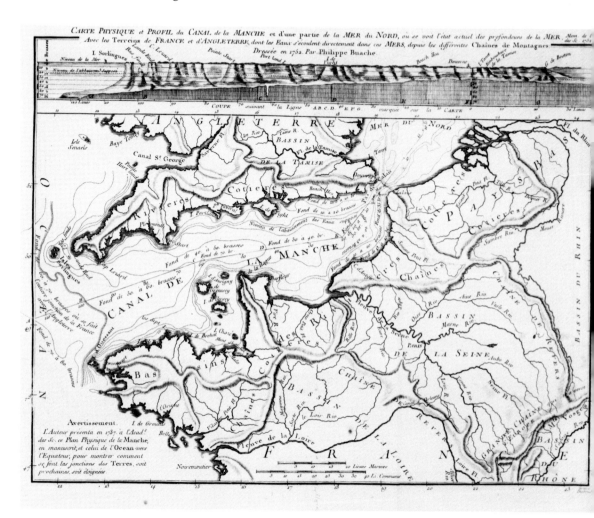

Figure 35  Buache's 1752 "Carte Physique et Profil du Canal de la Manche... où se voit l'état actuel des profondeurs de la Mer." Original 326 × 257 mm. Copperplate engraving. (Courtesy of the University of Wisconsin Memorial Library.)

After the mid-1700s geological mapping became more and more common. In 1762 G. C. Füchsel, another of the physicians with which early cartography seems to abound, combined his vocation with a geological avocation and made a map of a large part of Germany.[166] The first map to use color to distinguish the various formations appears to have been made in 1774 by F. G. Gläser, who observed that it would have been easy to give a more detailed written account but that he kept it brief because the reader could extend it by reference to his map.[167]

A considerably better colored map was published in 1778 by Charpentier, who pointed out in the foreword to his book that he had taken the map from his larger mineralogic map of Saxony, which he thought was a better map, but which was too large to bind in the book.[168] The map employs eight colors together with symbols for areas where the predominant rocks are granite, gneiss, slate, limestone, gypsum, sandstone, and so forth, together with other symbols for mineral types. Barometric observations provided some elevations based upon Wittenberg.

The first geological atlas, begun by Guettard and finished by Monnet in 1780 using a reduction of the Cassini topographical map as a base, has already been referred to in chapter 3. FIGURE 36 shows a portion of sheet 25 of the atlas.

Standard geological mapping remained much the same, but there was an increasing adoption of compilation and smaller scales for the thematic representation of geological phenomena. The 1811 map of the Paris Basin by Cuvier and Brongniart and W. Smith's detailed map of 1815, referred to in chapter 3, and most of the other geologic maps, such as the colored maps accompanying papers published by the Geological Society of London, were based on individual field observation. In the second decade of the nineteenth century compilation plus fieldwork began to produce smaller-scale geological maps, exemplified by d'Halloy (1813) of a large section of France, Greenough (1819) of England and Wales, von Buch (1826) of Germany, and Beaumont and Dufrenoy (1840) of France.[169] The first map of all Europe is probably that which accompanied Lyell's *Principles of Geology* (1834), and a geological map of the world was produced in 1843 by Ami Boué, another cartographer-natural historian trained as a physician. Berghaus's *Physikalischer Atlas* included fifteen geological maps, and Johnston's *The Physical Atlas* contained a geological world map.

A notable, clearly thematic, different kind of geological map was prepared by George Poulett Scrope in 1825 (FIG. 37).[170] Scrope, a student of natural history, had done much work in the volcanic area of central France,[171] but he was also interested in the worldwide distribution of volcanoes. His map was intended to illustrate, or at least lead toward, a theory of crustal deformation based on the geographical coincidence of volcanic eruptions and mountain ranges. The map employs a dark, red brown to show "Points of Eruption" and bands of dark gray to show "Lines of Elevation," that is, mountain ranges. Maps showing similar data were relatively common by midcentury. In later years Scrope turned his attention increasingly toward political economy, and, as we shall see in chapter 5, he also merits a "first," albeit a very small one, in the thematic cartography associated with that interest.

**Figure 36** Part of sheet no. 25, engraved by Dupain-Triel and dated 1766, in Guettard and Monnet, *Atlas et description minéralogique de la France*. Reduced 50 percent. Copperplate engraving. (Courtesy of the British Library.)

## THE LAND SURFACE

From the earliest days of cartography the existence of mountains had been shown on maps, but the representation of the form of

Maps of the Physical World 93

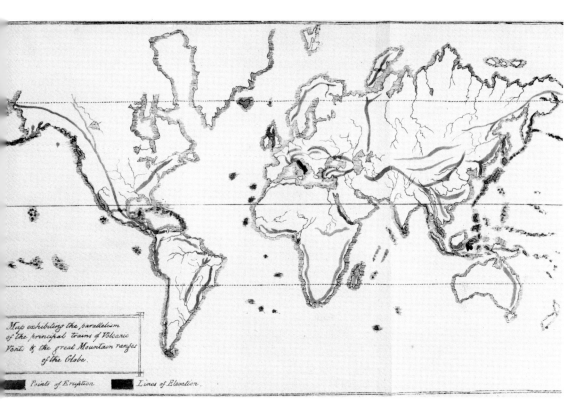

**Figure 37** Scrope's 1825 "Map exhibiting the parallelism of the principal trains of Volcanic Vents & the great Mountain ranges of the Globe." Original 335 × 215 mm. Lithograph, hand colored. (Courtesy of the British Library.)

the land in anything more than a cursory, diagrammatic fashion was a long time in coming. As Dainville wrote:

> The better cartographers at the end of the eighteenth century had no concern with the accurate portrayal of relief, while their successors at the beginning of the nineteenth century were greatly concerned with it.[172]

A successful display of the characteristics of the form of the land surface depends on the accurate reckoning of elevations together with a technique for portraying the third dimension. In the latter part of the nineteenth century neither was generally available, and even Cassini, in answering a complaint about the poor relief portrayal of Brittany on the "Carte de Cassini," wrote that accurate terrain representation was too large, difficult, and costly an undertaking. But attitudes can change quickly, and by the first part of the nineteenth century there was considerable interest in the relative elevations of the land surface. This came about for a variety of reasons, many of which were touched upon in chapter 2, such as canal building, interest in various interrelationships among elements of physical geography, and the fact that the military needed altimetric data for fortification plans, artillery fire, and troop cover.

By the end of the eighteenth century two methods of portraying the land surface had been introduced, and the means had been devised for obtaining altitudes for mapping. The contour, as Dainville so expressively put it, "had risen out of the sea like Aphrodite," and the precise hachure had been devised.[173] Only after the variables involved in barometric measure had been ascertained, particularly by Laplace, could that instrument be used with reasonable results. It was much easier to drop a weight on a line below the ready reference surface of the water than it was to determine relative heights on the land.[174] Consequently not much was known about heights at the beginning of the nineteenth century, and Humboldt is reported to have observed in 1807 that the height of only sixty-two mountains had been measured and that half of those had been accomplished by him.[175]

The portrayal of major and minor land features can be undertaken at all scales, and, if they are delineated so as to show the structural relationships of the three-dimensional forms, they qualify to be included in thematic cartography no matter what the scale. Nevertheless, until recent times much large-scale representation has been general in the sense that one could read or interpolate elevations but not get much of the effect of what in the German language is called *plastik,* a molded or modeled appearance. There were some notable exceptions, of course, such as the Dufour map of Switzerland and a few others.[176] This is not the place to review the very large subject of the development of the portrayal of the land surface, but it is appropriate to look at the developments in small-scale thematic cartography that led to the inclusion of relief maps in the two great thematic atlases, the *Physikalischer Atlas* and *The Physical Atlas.*

The story essentially begins in the early 1780s with the appearance of a small volume by Du Carla on the subject of using contours to show the form of the land surface.[177] In 1791 Dupain-Triel published a contour map of all of France, which was neither very accurate, because of the lack of data at that time, nor very expressive, since it had only a few contours (FIG. 38).[178] In 1798/99 (year VII) he published a map with the same contours but which was a good deal more expressive (FIG. 109). In addition to showing the "backbone" of uplands with a kind of sinuous

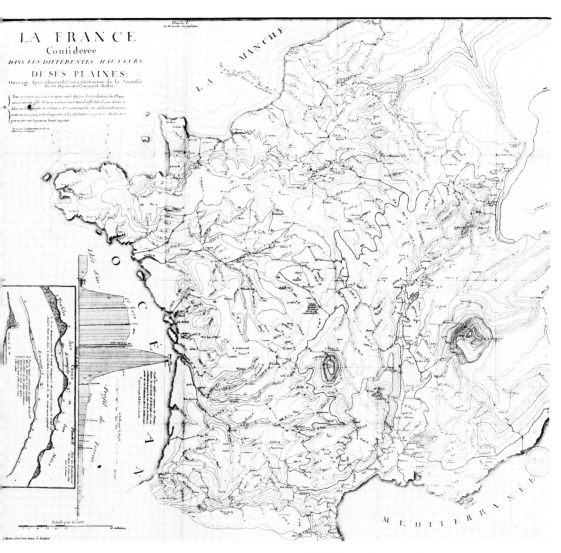

**Figure 38** Dupain-Triel's 1791 "La France considérée dans les différentes hauteurs de ses plaines." Original 555 × 495 mm. Copperplate engraving. (Photo. Bibliothèque Nationale, Paris.)

hachuring, the elevations were tinted in several sepia tones with the lowlands darker and the uplands lighter, a prototype layer-tinted map. Layer tinting, by employing a succession of tones or hues for categories of elevation, is clearly a thematic technique, since its objective is to enhance the representation of form rather than simply elevational data.

Published attempts to show the land surface structure of Europe thereafter were not very successful. One of the earlier ones is the attempt by Ritter to portray the surface of Europe in his 1806 atlas *Sechs Karten von Europa* (FIG. 39).[179] Maps using hachures were hardly more successful, for they were unable to convey an impression of relative height. What was wanted was the kind of thing Ritter had attempted combined with elevational information. The incentive was provided by the Société de Géographie de Paris, which had been founded in 1821.

**Figure 39** Ritter's 1806 "Oberfläche von Europa als ein Bas-Relief dargestellt," in *Sechs Karten von Europa*. Here title missing at top. Original 200 × 180 mm. Copperplate engraving, aquatint. (Courtesy of the Royal Library, Copenhagen.)

Immediately after its organization the Société developed a program of competitions and prizes to expand geographical knowledge. These ranged over a broad field as time went on, but, significantly, the first of its competitions had to do with the mountains of Europe. Announced in its *Bulletin* in 1822 it set a deadline of 1 January 1825.[180] Although no map was called for in the announcement, the second entry was accompanied by a manuscript map submitted by two Danes, O. N. Olsen and J. H. Bredsdorff (see FIG. 110).[181] The judges awarded them a half-prize in the form of a six-hundred-franc gold medal. A year later Olsen and Bredsdorff asked for and received a tracing of the map to use as a basis for the compilation of a more detailed map.

The new map and an accompanying *Commentaire* appeared in 1833 (FIG. 40).[182] The work had been announced for 1829–30, but soon after getting under way on the map Olsen's co-worker, Bredsdorff, was named professor at the Academy of Soröe, and Olsen was left to do the work himself. His nomination to the post of lecturer in topography at the military school in Copenhagen further delayed matters. Olsen reported in the *Commentaire* that he had some twenty thousand elevations to work with, a far cry from the situation twenty years earlier. This made it possible to use contours, instead of the hachures on the 1824 manuscript map, and Olsen stated that he was unaware of anyone having earlier tried to do this for a large area.[183]

There are two states of the copperplate of the map, a first one with contours and names of land features, and a second with additional names of land features, names of seawater bodies, and added hachures. The second state, normally colored orographically—that is, with colors used to delimit the mountain systems listed at the left and upper portion of the map—is shown uncolored in FIGURE 40. Two colored versions of the first state were distributed, one with geognostic coloring (five rock types, being a kind of generalized geologic map) and the other with hydrographic coloring outlining drainage basins and divides. Olsen's contour map was very influential for some twenty years, as we shall see presently. He made the right thematic contour map at the right time.

Oluf Nikolay Olsen was born in 1794, and in 1808 was appointed a cadet at the Artillery Institute of Copenhagen.[184] In 1812 he became a second lieutenant, and he passed his surveyor's examination in 1815. During the winters of 1817 and 1818 he studied drawing at the Academy of Fine Arts and had private instruction in painting. He taught surveying and drawing at the Artillery Institute and was also tutor in topography for the crown prince. He was active in the official surveys of both Denmark and Iceland, taught topography in the Royal Military Academy, and rose to be in charge of the topographical section of the General Staff. As we already have seen, he collaborated as cartographer with Schouw, the botanist.

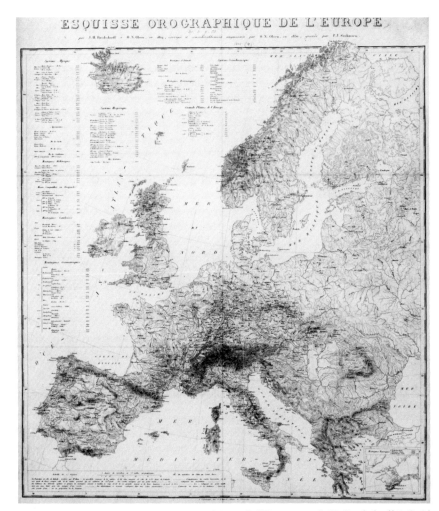

**Figure 40** Olsen's 1833 "Esquisse orographique de l'Europe par J. H. Bredsdorff & O. N. Olsen, in 1824; corrigée et considérablement augmentée par O. N. Olsen, en 1830...." Original 540 × 623 mm (1:6,543,100). Copperplate engraving. (Courtesy of the British Library.)

Berghaus recounted that he had made a physical map of France in 1821, and that since then he had been collecting data to make a better map of Germany and adjacent areas.[185] Before Berghaus began work on the new map, Olsen sent him a copy of his work, and Berghaus thought so highly of it that he essentially copied it for the 1842 map of Europe's major mountain systems in the *Physikalischer Atlas*. Berghaus's map was of a more limited area and incorporated some new data for the Balkans and Sardinia. Johnston in turn copied Olsen's and Berghaus's map for a somewhat larger area in his *The Physical Atlas*. The details of

Maps of the Physical World

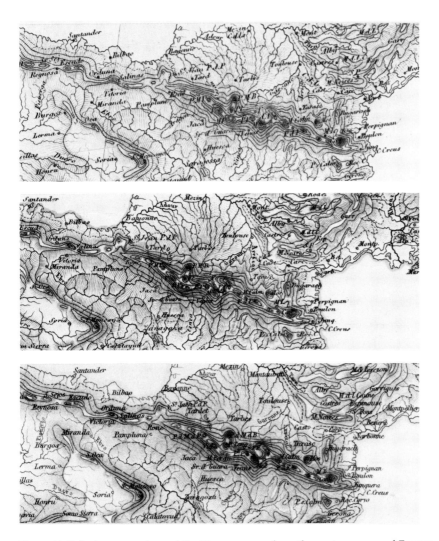

**Figure 41** Full-size comparison of the Pyrenees area from the contour maps of Europe. *Top:* Olsen-Bredsdorff map (1833); *middle:* Berghaus's *Physikalischer Atlas*, third Abteilung, Geology, no. 3 (1842); *bottom:* Johnston's *The Physical Atlas*, Geology, no. 5 (1848). (Courtesy of the British Library.)

the three maps are almost identical, as FIGURE 41 shows. The great value of Olsen's thematic contour treatment of the "Mountains of Europe" was that through its reproduction in two widely used atlases it, in a sense, popularized the contour as a device for showing absolute and relative elevations with a degree of accuracy which did not depend upon cluttering the map with spot heights. It was a method ideally suited to coping with the data revolution of the time.

Even before Berghaus's *Physikalischer Atlas* appeared, other small-scale thematic treatments of the land surface were appearing. Emil von Sydow produced wall maps of each continent for schools in 1838–40 showing form with hachures and with three layer tints, and these were followed by a series of school atlases. The cartography in these was considerably more thematic, as shown by FIGURE 42. From about 1840 on the thematic treatment of the land surface was common but often imaginative.[186] One of the more interesting treatments is that in A. K. Johnston's 1852 atlas in which six of the eighteen maps are of the landforms of the continents and the British Isles. FIGURE 43 is an example.

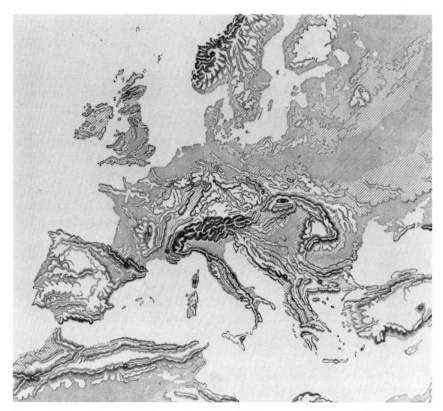

**Figure 42** A portion of the brown layer tint for the map of Europe in Sydow's 1847 *Schulatlas* as included in his 1855 *Orographischer Atlas*. (Courtesy of the British Library.)

### VEGETATION AND ANIMAL LIFE

Forests, various kinds of plants, and pictures of animals have shown up on maps almost from the beginning, and certainly by the latter half of the eighteenth century wooded areas were a regular part of the large-scale, topographic maps then becoming more and more avail-

Maps of the Physical World

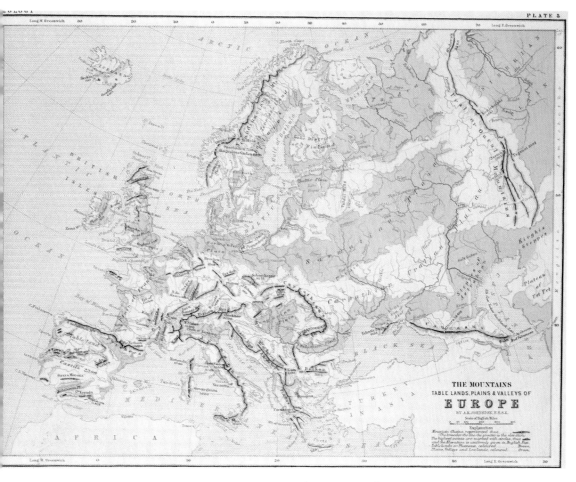

**Figure 43** "The Mountains, Table Lands, Plains & Valleys of Europe" in Johnston's 1852 *Atlas of Physical Geography*. The broader the line showing mountains, the greater the elevation, with the highest points marked with a circle. Tablelands, brown; plains, valleys, and lowlands, green. Original 308 × 236 mm. Lithographic engraving, printed color. (Courtesy of the British Library.)

able, but the systematic mapping of plant and animal life essentially begins with E. A. W. Zimmermann's map of animal distribution in 1777 and Ritter's innovative atlas *Sechs Karten von Europa*, the maps for which were prepared from 1804 to 1806. In addition to his *Bas-Relief* (FIG. 39), Ritter included a map showing the distribution of cultivated plants, one of wild trees and shrubs (FIG. 19), and one of tame and wild mammals. Like Zimmermann's map, these are merely outline maps on which the various names of the items are lettered here and there and on which boundaries of some species are noted. Latitude lines are prominent to emphasize the climatic relationships. In addition to the *Bas-Relief*, Ritter also included a map tracing the crests of the mountain ranges so their relations to the distributions could be appreciated.

The pictorial profile, sometimes relating elevation to distributions, especially vegetation, had already become rather common in the early part of the nineteenth century, and it gained in popularity as time went on.[187] These displays became very elaborate and were published both separately and in atlases; but they were not maps.

The first real attempt to map the distribution of vegetation was undertaken by J. F. Schouw, a Danish botanist, who published a book on the fundamentals of plant geography in 1823 and accompanied it with an atlas of twelve maps.[188] In the foreword to the book he credits Humboldt and Wahlenberg for his initial interest in plant geography, which was heightened by a mountain trip to Norway and the opportunity to study plant geography in southern Europe. Both the book and the atlas are pioneering attempts to describe and portray the vegetative cover of the earth.

The atlas consists of twelve sets of paired eastern and western hemispheres on facing pages, showing by colors the distributions of various classes of plants, such as cereals or palms (FIG. 44). Each map is briefly described in the atlas, and the user is referred to the appropriate pages in the book where the distributions are discussed. Schouw well understood both the advantages of a map and its limitations. He cautions the reader about the cereals map as follows:

> This plate shows in which parts of the earth cereals are cultivated...and the species of cereals. The five realms, shown by various colors, are not so sharply bounded in nature as on the map, and the cereal species signified is usually not exclusive but only predominant. Thus, other species than rye are cultivated in the parts of the earth colored yellow, yet there rye provides the main foodstuff from the vegetable kingdom. Where another, besides the predominant cereal, is common enough to be nearly equal to it, this is indicated by lines of the corresponding color, e.g., in the coastal lands of the Mediterranean Sea, where wheat no doubt predominates, but maize, and in some areas rice, are also cultivated.

The last map in the atlas is an attempt to divide the earth's vegetative cover into plant regions by using colors and names for regions. Cartographically Schouw's maps are unusual for the time, in that color carries the message rather than being employed merely as an enhancement of a monochromatic portrayal.[189]

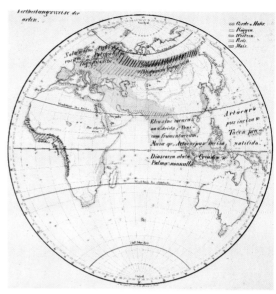

**Figure 44** Map 5 showing the diffusion districts and distribution of kinds of cereals from Schouw's 1823 *Pflanzengeographischer Atlas*.... Original hemispheres 353 mm diameter. Base map lithographic engraving, labeling overprinted, hand colored. (Courtesy of the British Library.)

In 1833 Schouw published in Copenhagen a physical-geographical description of Europe and accompanied it with an atlas.[190] In addition to physical maps of landforms, and temperatures, two other maps show the distribution of wild trees and shrubs and the important cultivated plants. The maps are reminiscent of Ritter's simple maps, but they are much more sophisticated and show the northern boundaries of seven food plants, such as maize, fruit trees, and wheat.

Berghaus's *Physikalischer Atlas* contains six maps on botanical geographical subjects, and by 1839 such maps had become quite complex (FIG. 45). Included was a map of Europe presenting Schouw's distribution of products, limits of growth, isotherms, and so forth. Johnston's *The Physical Atlas* contains two similar maps including one showing Schouw's world plant regions (FIG. 46). The cartographic techniques used for the thematic mapping of natural vegetation and cultivated plants were basically straightforward, employing color and patterns as area symbols, lines as boundaries of extent and isolines, and variable shading to show variations in concentration. From the 1840s on the mapmaker's problem became one of acquiring sufficient data, not of how to portray it. These first maps of vegetation were relatively crude, but as the data improved there was a rapid development in cartographic quality.[191]

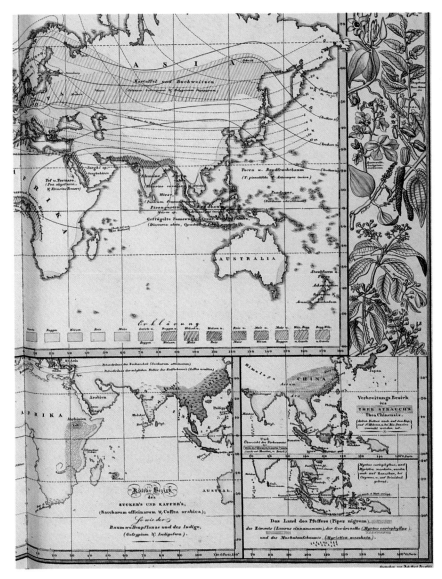

**Figure 45** Part of the 1839 world map of cultivated plants supplying main foodstuffs in Berghaus's *Physikalischer Atlas*, fifth Abteilung, Botanical Geography, no. 2. Reduced 48 percent. Copperplate engraving, hand colored. (Courtesy of the British Library.)

The first maps of animal distribution appear to have been made somewhat earlier than those of vegetation. Although Eckert refers to the 1783 map of quadrupeds by Zimmermann as being the first map of animal geography,[192] Zimmermann included an earlier version in a

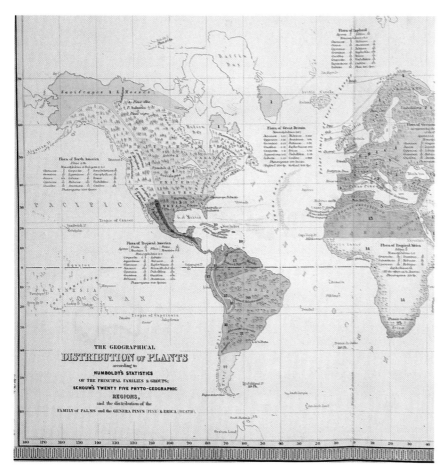

**Figure 46** Part of the world map of the distribution of plants, showing Schouw's twenty-five phytogeographic regions, in Johnston's 1848 *The Physical Atlas*, Phytology and Zoology, no. 1. Reduced 60 percent. Copperplate engraving, hand colored. (Courtesy of the British Library.)

book published in 1777. Both merely have names scattered over the map along with some limits of distribution. Ritter's *Sechs Karten von Europa* also contains a primitive map of tame and wild mammals with names here and there and with some limits shown. None of these maps qualifies as a good example of thematic cartography, since they do not portray distribution very effectively. Not surprisingly, the clear display of distribution and variation starts with Berghaus's *Physikalischer Atlas*, which included twelve zoological maps mostly dated 1845. Johnston's *The Physical Atlas* combined some of these and included five plates showing with many maps the distribution of many kinds of birds and animals, ranging from monkeys to carnivores, and from rodents to reptiles (FIGS. 47 and 48). The variable density of occurrence is shown by shading.

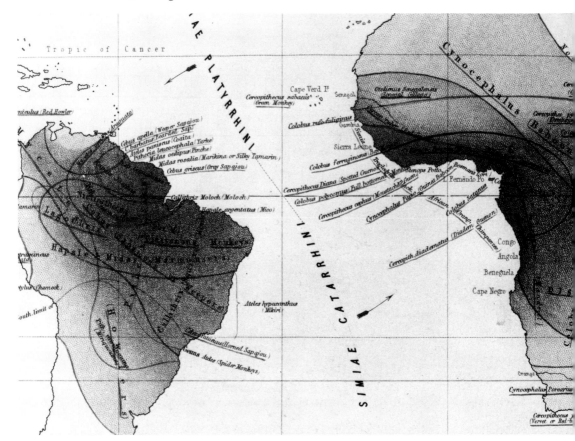

**Figure 47** Part of the world map of the distribution of monkeys and lemurs in Johnston's 1848 *The Physical Atlas*, Phytology and Zoology, no. 3. The division between the New World *(Platyrrhini)* and Old World *(Catarrhini)* monkeys is shown by the diagonal lettering and the arrows. About full size. Copperplate engraving, hand colored. (Courtesy of the British Library.)

In the two centuries following Kircher's first attempt at mapping a physical phenomenon (ocean currents), the methods of displaying the occurrence of the facts of natural history had improved considerably, but the basic objective behind Kircher's map and Halley's maps of winds and compass declination was not to change. To portray the variety of earth distributions so that their character could be seen "at a glance" became a necessity; descriptions in words simply could not suffice.

Maps of the Physical World

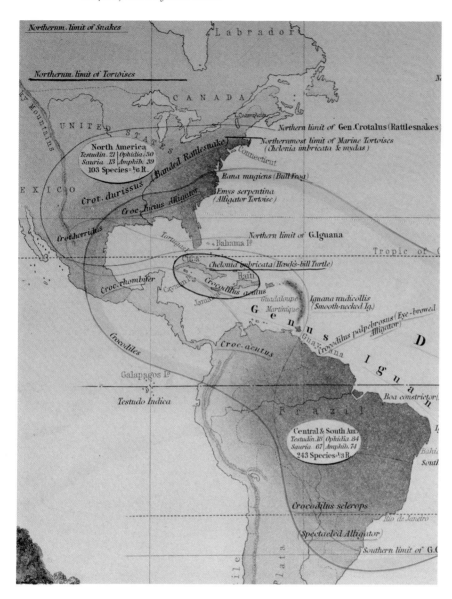

**Figure 48** Part of the world map of the distribution of serpents in Johnston's 1848 *The Physical Atlas*, Phytology and Zoology, no. 7. About full size. Copperplate engraving, hand colored. (Courtesy of the British Library.)

As one reviews the early development of thematic cartography of the physical world up to the mid-nineteenth century, the symbology seems to have kept up with the proliferation of data. As soon as the data were available, a method of symbolizing it appeared on the scene. If Humboldt had not applied the Halleyan line to temperature data, someone else would have. Probably the greatest difficulty lay in the use of color. It was obviously suited to showing distributions of anything, such as rocks or plants, that varied in character over area, but no mechanical way of applying it was generally available until after the mid-nineteenth century. The widespread practice of hand coloring each individual map, slow and expensive, is a measure of the intellectual demand of the time for the graphic portrayal of geographical data on thematic maps.

# *Five*

# MAPS OF PEOPLE
# AND THEIR ACTIVITIES

THE THEMATIC MAPPING OF THE WORLD OF HUMANITY DID not parallel the cartographic portrayal of the physical world; instead, it lagged by more than a century. The rise of the physiocrat school in France in the eighteenth century, with its espousal of laissez-faire, the publication in 1776 of Adam Smith's *An Inquiry into the Nature and Causes of the Wealth of Nations,* the developing industrial revolution with its accompanying rapid urban growth, and the general increase in population in western Europe all combined to foster an increased interest in the central element—people. To be sure, such things as life expectancy had been studied earlier, and the term "political arithmetic" had come into use; but any systematic, broad inquiry into people and their activities required adequate statistics, and these did not become generally available until after the beginning of the nineteenth century. A few maps, more general than thematic, of nonphysical subjects did appear in the eighteenth century, several of which have been noted in chapter 3. Hensel's maps of languages and Crome's product map of Europe are conspicuous because they are so alone.[193]

From a technical point of view the attempt to portray geographical structures of the distribution of people, their activities and conditions, raised some new cartographic problems. Foremost was the fact that much of the data consisted of summary statistics relating to areas, such as census enumeration districts, *départements,* or even whole countries. The problem of dealing with such numbers on maps had not been faced before. It was obvious to anyone who thought about it that, all else being equal, anything counted in large areas would total more than the same thing counted in small areas. Since the primary objective of a thematic map was usually comparison among regions, this necessitated removing the effect of the variations among the sizes of the districts. This led to the introduction of the concept called relative or specific, now called density, which refers to some phenomenon, such as population, in which the enumerated total is divided by the area of the enumeration unit, resulting in a number of units per square kilometer or square mile. Various other ways of making the statistics comparable were tried, such as the reverse of the foregoing, yielding a statistic such as the number of square kilometers or square miles per unit of the phenomenon.[194] Another common solution was to derive departures from an average.

One other potential technical problem, variable shading, was solved, or nearly solved, about the same time that this class of thematic maps began to be made. Most of the characteristics of natural history that had been portrayed on maps before the 1820s, such as geological, geomagnetic, or meteorological phenomena, required only nominal differentiation by colors or quantitative portrayal by isolines. Variable shading, such as that which might be used to make a layer tint map or a "plastic" shaded relief map, was both difficult and expensive to produce on a copperplate, though it was done, as for example on Dupain-Triel's contoured "Carte de la France..." (FIG. 38) and Ritter's "Oberfläche von Europa..." (FIG. 39), as well as other shaded relief maps of the early nineteenth century. Both continuous (variable) tone and flat (uniform) tone were forms of representation required by these new kinds of thematic maps, and fortunately both were possible in copperplate engraving and lithography. More on these methods can be found in chapter 7. The introduction of lithography proved to be a distinct advantage to the new class of thematic mapping, primarily because it was cheaper than copperplate engraving. This meant that numerous thematic maps were made which likely never would have seen the light of day if copperplate engraving had been required. Lithography was not indispensable, of course, and if it had not come on the scene for another half-century thematic maps would still have been made; but they probably would have been far fewer. The availability of lithography certainly promoted trying out ways of representing these new kinds of data.

In this chapter and the next I will review the earlier attempts to map various kinds of statistical data referring to geographical areas, but before doing so it will be helpful to note the basic ways by which these kinds of data can be presented cartographically. There are five ways of mapping such statistics. First, one may simply write numbers directly on the map in the appropriate enumeration units, such as counties or *départements.* Second, the totals for the areas may all be divided by some unit value, and then some marks, such as uniform dots, may be placed in appropriate positions in each area, each mark then representing so many of the things being mapped. The variation in the numerousness of the dots from place to place on such a "dot map" is intended to portray the different densities. A smoother effect can be obtained by employing continuous tone shading to provide the same visual effect. A third method involves employing some kind of mark for each enumeration, the size of each mark being varied in proportion to the number it represents. Although other shapes can be used, circles were, and still are, commonly employed because they are easier to draw. A fourth method first requires the calculation of some kind of comparable number for each enumeration unit, such as a density, a ratio, or a departure from an average. The range of such derived numbers is then usually divided into several categories which are then symbolized by flat tones, patterns, or colors. This has come to be called the choropleth method. A fifth method is quite different. In this the density numbers or ratios, when plotted in their enumeration units on the map, are conceived as being like spot heights or temperatures, among which isolines (often called isopleths) may be interpolated, just as one would draw contours or isotherms. The ingenuity of cartographers is such that a good many refinements or variations of these basic systems have been devised.

In the next two chapters we shall focus on a variety of categories of thematic maps in which man, his activities, and his social environment were the central themes. Needless to say, these cover broad areas, and since space is limited I must be selective and focus on those elements that seem to have been of major significance in the development of thematic cartography. The order seems immaterial except that I shall start with the portrayal of the simplest basic statistics, numbers of people.

POPULATION

Many early maps implied variations in population, at least of cities and towns, by using little drawings and other marks of different sizes, but the portrayal of the populations of countries or of parts of countries came rather late. One of the earlier, perhaps the earliest, is Ritter's map no. 6, "Tafel über die Bewohner von Europa über Volksmenge und Bevölkerung dieses Erdtheils" in his atlas *Sechs Karten von Europa,* in which population numbers and densities are simply written in on the map.[195]

Another early example of the practice of putting population numbers on a map is J. Wyld's 1815 map "Chart of the World Shewing the Religion, Population and Civilization of Each Country" (FIG. 104).[196] It is likely that there are other such maps on which numerals have been written, but they are of more concern to those tracing the evolution of demographic interest than to historians of cartography. Numerals, of population or anything else, in some instances may be more useful when written on a map than when arrayed in a table, but such maps hardly qualify as thematic since they do not graphically portray the character of a distribution at all effectively.

The earliest simple dot map of population appears to be one by Frère de Montizon, dated 1830, in Paris (FIG. 49).[197] It is based on *département* data with each dot representing 10,000 people *(têtes)*, and the dots are irregularly spaced within the *départements*. The map is framed by a table listing map numbers, names of *départements*, their populations and chief towns, together with a variety of philosophical comments and quotations. Two states of the map are known, printed from two different plates.[198]

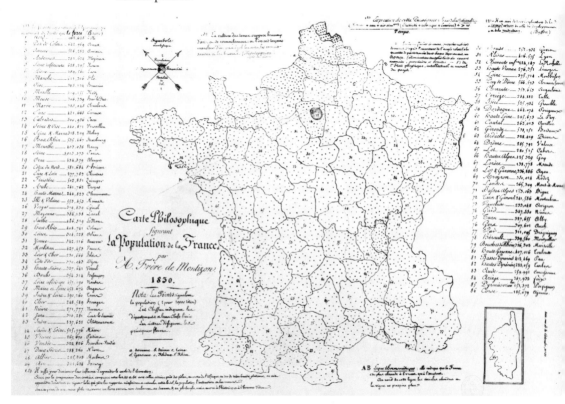

**Figure 49** Frère de Montizon's map of the population of France, dated 1830, plate 1. Original, including marginal tables, 619 × 478 mm. Lithograph. (Photo. Bibliothèque Nationale, Paris.)

Frère de Montizon is a mystery man who merely listed himself on the map as *"Professeur de Chimie, de physique et de mathématiques, nommé par le Conseil royale de l'Instruction publique."* The National Archives does indeed list a person by that name, but that is all that can be ascertained of his qualifications. It is probable that he had a printing shop located in various places in Paris between 1823 and 1837, and it is known that he wrote numerous educational items on history and philosophy directed primarily toward the young.[199]

It is one of the accidents of history that Frère de Montizon's invention of the thematic dot map should have gone completely unnoticed. In terms of cartographic innovation it ranks with the isothermal map, yet as far as can be ascertained no reference to him or to his dot map was made by anyone well into the twentieth century.[200] Frère de Montizon's dot map is conceptually much more of an innovation than is the isotherm, which is but a short step from the curve lines of geomagnetic phenomena, and it is a reasonable supposition that if Humboldt had devised the dot map the innovation would have been heralded and quickly put to use by others. Instead, except for a few rather crude, large-scale applications, without clear unit values, in medical mapping (see chap. 6) we will see in chapter 7 that this basically simple, logical idea had to wait some thirty years to be reinvented and much longer than that to become generally known.

The concept of population density had been well understood for a long time, but except for Ritter's simple density numbers on a map in *Sechs Karten von Europa,* no other map of population density seems to have appeared until 1828. In that year a remarkably sophisticated, manuscriptlike *Administrativ-Statistischer Atlas vom Preussischen Staate* appeared. Hand drawn on a copperplate engraved base map, less than a hundred copies seem to have been produced. One of the twenty-two maps in the atlas is a colored choropleth map of population density (FIG. 50). The author of the thematic component is not known, but may have been C. von Rau.[201]

The next choropleth population map appeared in 1833 as a frontispiece in a book on political economy.[202] Its author was Scrope, the geologist referred to in the previous chapter who also later became interested in political economy. The map is small and crude, but it is apparently the first world map to portray population density directly by dividing the earth into regions "averaging more than 200 inhabitants to the square mile" (black), "from ten to 200 inhabitants to the square mile" (gray), and "surfaces averaging less than ten inhabitants & rarely so much as one to the square mile" (blank). There is not a word about the map in the book, but in a second edition published forty years later a redrawn but little-changed map was again included as a frontispiece, this time accompanied by a short, explanatory appendix pointing out the difficulty of deciding which areas, that is, states, countries, regions, or continents, to use as the basis for comparison.[203] Fundamentally, the technique Scrope employed is a sophisticated variant of the choropleth method, today called "dasymetric," which was to be employed in careful fashion several years later and which will be described below.

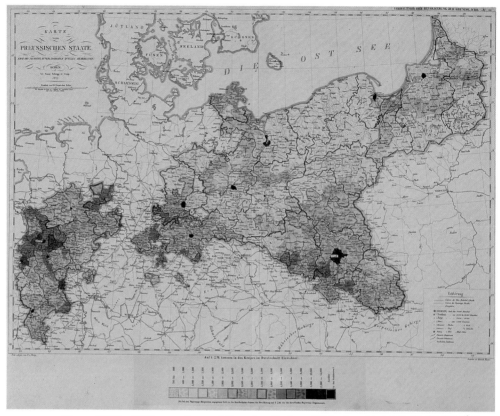

**Figure 50** The 1828 map of population density (no. 13) in the *Administrativ-Statistischer Atlas vom Preussischen Staate*. Original (map) 701 × 488 mm. Base map copperplate engraving, thematic data hand colored. (Courtesy Geheimes Staatsarchiv Preussischer Kulturbesitz.)

Scrope clearly gave the matter little thought and did nothing more than make a simple, rough sketch map, but this was the first of several similar small, world maps. For example, Berghaus included such a map of population density as part of the *Anthropographie* section of the *Physikalischer Atlas* (FIG. 51), and Petermann made one in 1859.[204]

Scrope, born George J. P. Thomson in 1797, assumed his wife's name in 1821. He was a keen geologist whose studies on volcanoes and ideas on geologic processes were authoritative and advanced; he became a member of the Geological Society in 1824 and the Royal Society in 1826. He had a lively interest in politics as well and served for more than thirty years in Parliament as an advocate of free trade and especially of social reform. It is not surprising that the notion of displaying population density on a map occurred to him, for he had used maps as a

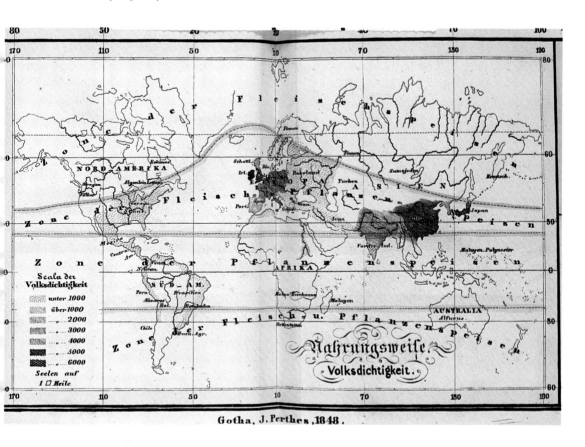

**Figure 51** The 1848 world map of population density and food habits in Berghaus's *Physikalischer Atlas*, seventh Abteilung, Anthropography, no. 1. About full size. Copperplate engraving, hand colored. (Courtesy of the British Library.)

geologist, and as previously noted he had made them to accompany his geological books of 1825 and 1827.

A straightforward choropleth map of the density of population by *départements* in France was published in 1836 by d'Angeville (FIG. 52).[205] The density values were calculated for square *myriamètres* (100 square kilometers), and the date of the data is given on the map as being 1831. The *départements* are listed in rank order in a table adjacent to the map of France. The rank of each is given on the map, and the total list, less Corsica, is divided in five series of seventeen *départements* each. Each series is assigned a textured darkness ranging from white for the greatest population density to the darkest for the lowest, the reverse of the usual arrangement of the tonal scale. The range is obtained by a series of linings and crosshatching.

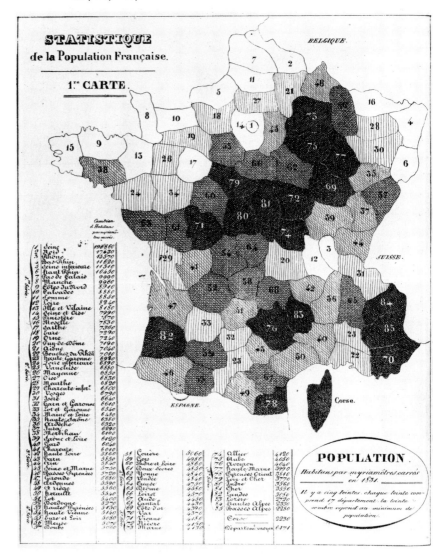

**Figure 52** D'Angeville's 1836 map of the number of persons per square *myriamètre* in France. Original 187 × 239 mm. Lithograph. (Photo. Bibliothèque Nationale, Paris.)

The population density map by d'Angeville is one of sixteen choropleth maps he published in his essay on the population statistics of France, and others will be noted in chapter 6. He argues in favor of the graphic method, pointing out that, even though "speaking to the eyes" (following the example of Baron Charles Dupin) has been frowned upon by some others, it enables one to focus on the essential character of a

**Figure 53** Harness's 1837 map of the number of persons per square mile in Ireland. Original 1:633,600 (1 inch to 10 miles). Copperplate engraving, aquatint. (Author's collection.)

distribution, which is often hidden in the "aridity of the enumerations."[206] As I shall note in chapter 6, the Baron Charles Dupin to whom he referred appears to have been the first to make a choropleth map, ten years earlier.

The remarkable development of the thematic mapping of population in the 1830s continued with yet another, far more sophisticated map published in 1837 (FIG. 53). It appeared in a publication which, unfortunately, had only limited distribution, *Atlas to Accompany Second Report of the Railway Commissioners, Ireland.*[207] As mentioned before, the development of the steam railway as a practicable means for moving goods and people was followed by a very rapid expansion of lines, with much attendant confusion about financing and the role of the state. To make the development of railways in Ireland more orderly, a commission was appointed, and to illustrate their second report they caused the *Atlas* to be prepared.

In the history of thematic cartography the *Atlas* shines as bright as the limelight invented by one of the commissioners, Thomas Drummond, since it put to use no less than four technical innovations, not at all in rudimentary fashion, but in finished, sophisticated form, accompanied by detailed written explanations. Because Ireland was not of great general interest at the time, and this being just another government report, and not having wide distribution, the innovative maps escaped much notice. As is not unusual, the techniques were used by others without credit or had to be reinvented. Except for the geological map by Richard Griffith, the *Atlas* was noted only briefly upon its appearance and then was soon forgotten for more than a century.[208]

The author of the innovative maps, Henry Drury Harness, was a young (born 1804) lieutenant in the Royal Engineers who, except for military survey training, had no experience in mapmaking, especially thematic. At that time very few people had had any such experience with respect to mapping statistical data, and the lack of such background may have been an advantage, since neither he nor his superiors had been conditioned to any preconceived methods. At any rate, the population map is a model of cartographic procedure. It employs a refinement of the choropleth method, it separates rural and urban populations, and it portrays the former by shading and the latter by proportional circles. A clear understanding of the objectives and procedures is indicated by the following explanation by Harness:[209]

> The first map is intended to convey, at once, to the mind, an idea of the manner in which the population is distributed over the country, and for this purpose the depth of shade applied to each part has been regulated by the density of the population.... From these data the number of inhabitants, per square mile, in each barony was determined, after deducting the population of the towns intended to be shown upon the Map, and the relative depth of shade for each barony was expressed by the numbers thus obtained. The towns are represented by dark spots, of which the areas are regulated by the number of their inhabitants.
> It was at first proposed to abide throughout by the

principle just described; but the Commissioners having expressed a wish that something more detailed than shading, by baronies according to their average population, should be attempted, they marked upon a map, such barren tracts of bog and mountain they knew to contain but few inhabitants; assigning also, from their general acquaintance with the country, what they supposed might be assumed as the population, per square mile, in such parts. The population of each barony, containing a portion of these boggy and mountainous tracts, was therefore considered, after allowing a few inhabitants to the barren parts, to inhabit the remaining portion, and the population deduced for that part, from the data thus furnished.

It is evident that the boundaries of the mapped classes are not those of the enumeration units (baronies), as would be the case with a simple choropleth map, but instead are more meaningful boundaries located where densities are assumed to change, a cartographic technique today called dasymetric.[210]

The darker the shading the greater the density, exactly opposite the system employed by d'Angeville. It was intended that there be four categories of shading, but the class limits were not stated, and the lack of control in the production of the copperplate engraving has, so far, made it impossible to determine them. The use of graduated circles to represent urban populations was not a new idea in statistical graphics (see chap. 7), but no earlier instance of their use on a map is now known. The Railway Commission population map is certainly remarkable, and whether credit for the ingenuity it reveals should go to Harness, to one of the commissioners, or to someone else we shall probably never be certain.[211]

The honor of being the first map of population density (persons per square mile) to be issued as part of an official census also belongs to Ireland. It appeared in 1843 when the Census of Ireland, 1841, was published (FIG. 54).[212] The commissioners explained:

We annex a map (Plate I) shaded so as to assist the eye in seizing a general view of the comparative density of the rural population. The figures under the towns indicate their population and the figures in the body of the map the population of the localities within which they are placed. The population of the towns (over 2,000) has been deducted before taking the averages, a precaution we thought necessary, though we believe it has not been usually done, so that the shading only exhibits the rural population.[213]

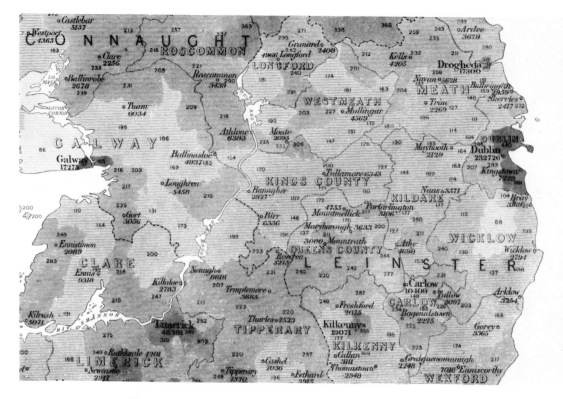

**Figure 54** Part of plate 1, "Population," showing the number of persons per square mile, in the *Census of Ireland, 1841* (published 1843). About full size. Copperplate engraving. (Courtesy of the British Library.)

In deducting the population of the towns, they followed the practice introduced on the Railway Commission's population map but stopped short of making it dasymetric. The census map is strictly choropleth, but, since the enumeration districts are small, a considerable amount of detail is incorporated.

Thomas Larcom, later Sir Thomas, was one of the three census commissioners and was no doubt responsible for the maps; he was a remarkable person.[214] Born in 1801, he joined the Corps of Royal Engineers in 1821 with a high reputation from the Royal Military Academy at Woolwich. He was posted to the Ordnance Survey in 1824 and was selected by its director, Colonel T. F. Colby, to go to Ireland where the survey was operating. In 1826 he was made assistant to Colby and was placed in charge of the Irish Survey, a responsibility which was to continue until the mid-1840s, since Colby was required in London.

Other population maps were soon to appear. In 1849 two population maps were made in London, one of the British Isles by August Petermann, a cartographer-geographer, and the other of England and Wales by Joseph Fletcher, a statistician. Their different viewpoints are reflected in the kinds of thematic maps they prepared. Fletcher's map, relatively small, shows the density of population by the proportion of inhabitants to 100 acres below (−) and above (+) the average of all England and Wales (FIG. 55).[215] A "Scale of Tints" obtained by lining and crosshatching symbolizes seven categories on the map with unusual class limits. The density was calculated for each county and for all of England and Wales, and then the percentage difference between each of the counties and the overall value was calculated. This determined the tint. The counties were arranged in rank order in an accompanying table, and the rank number of each county was entered on the map.

Fletcher's population density map, probably the first of England, was made to accompany eleven other maps of "moral statistics" which are listed in chapter 6. In the long paper in connection with which the maps appeared, Fletcher wrote:

> In all the Maps it will be observed that the *darker* tints
> and the *lower* numbers are appropriated to the
> *unfavourable* end of the scale. [Italics in original]

Of course, such a system does not fit a scale of population density, and he had to choose either high or low density as unfavorable. He chose high as darker, probably because of an assumed geographical correlation with moral statistics to which value judgments can properly be ascribed.

Fletcher's objective was to compare the geographical variations of diverse sets of statistics, and for that purpose choropleth county maps of departures from an average could be helpful. Petermann, on the other hand, as a cartographer-geographer, was more interested in the details of distribution, and he very possibly felt that the generalizing constraints of a choropleth portrayal were too much. Furthermore, he had just recently prepared a map showing the incidence of cholera in which he had employed a system of shading with the darker equated with more and the lighter with less. His training and experience had taught him to "look beyond" the relatively coarse information provided by county data, which homogenizes much geographical variation, and to try to portray on a map the true character of the distribution. If he had known about it, he could have employed Frère de Montizon's dot system (but using a better dot value and placing the dots more carefully), and were he living today that is probably what he would do. He did use the same cartographic technique he used on his earlier cholera map—variable shading—which if done well is more expressive of subtle variations than any other method (FIG. 56).

*Maps of People and Their Activities*

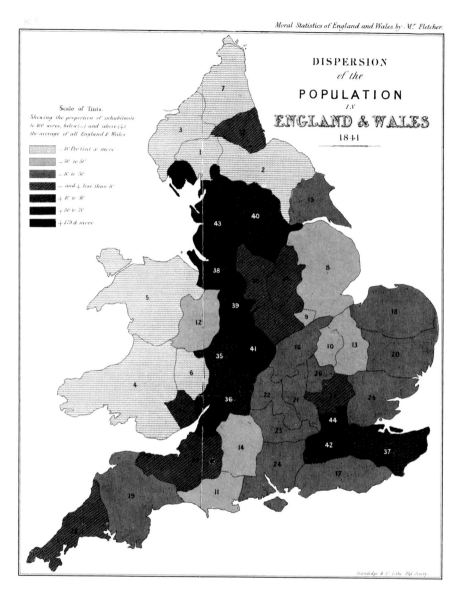

**Figure 55** Fletcher's 1849 "Dispersion of the Population in England & Wales." Original 190 × 235 mm. Lithograph. (Courtesy of the University of Wisconsin Memorial Library.)

The population map based on the census data of 1841 is an elaborate and carefully drawn portrayal of the details of distribution. Rivers and county boundaries are shown, and the calculated density values of number of persons per square mile are given on the map for each county. Cities and towns are shown by circles and area outlines, some

Maps of People and Their Activities

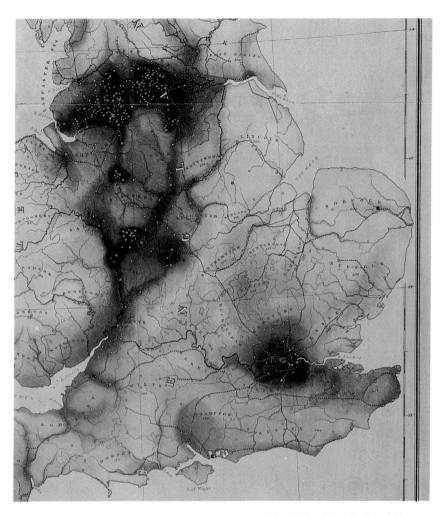

**Figure 56** Part of Petermann's 1849 population map of the British Isles. Reduced 50 percent. Copperplate engraving, aquatint, hand colored. (Courtesy of the British Library.)

white and others colored to indicate dominant occupations. The subtle shading shows the relative degree of concentration, the darker the greater. Although the handsome map was drawn by Petermann, the preparation of the copperplate engraving and the shading were expertly done by J. Dower.

Unlike many of the early makers of thematic maps, Petermann was not primarily a statistician, lawyer, doctor, physical scientist, or political economist who occasionally dabbled in cartography to satisfy his curiosity about geographical distributions. Instead, he was a cartographer-geographer whose education and training in his field was as good as was available at the time.[216] He was born in 1822 of poor parents in the little town of Bleicherode in central Germany, and at the age of seventeen he entered Berghaus's Engraver's School for Geography (*Geographische Kunstschule*) at Potsdam.[217] Petermann had no money and the state could not pay for his training, so Berghaus provided it free and even took the young man into his home, almost as a foster son. At the conclusion of his training, Petermann and his fellow student Heinrich Lange wanted to travel to Paris, London, and Edinburgh to widen their experience with geographical engraving. There being no official subsidy available for such travel, the two young cartographers took advantage of the opportunity to work on *The Physical Atlas* being prepared in Edinburgh by A. K. Johnston, based on Berghaus's *Physikalischer Atlas*. Petermann worked in Edinburgh for the period 1845 to 1847 primarily on the geologic and zoologic maps for the atlas. The maps for *The Physical Atlas* being completed, he moved to London in 1847, where he became active in the Royal Geographical Society, enhanced his lithographic skills, and worked as a geographical draftsman on maps for publication by himself as well as by others. In 1850 he published an *Atlas of Physical Geography* . . . and *A Descriptive Atlas of Astronomy, and of Physical and Political Geography* containing seventy-one quarto maps, thirteen being thematic physical maps of the kinds of topics included in the larger Berghaus's *Physikalischer Atlas* and Johnston's *The Physical Atlas*. He served as a committee member of Section F: Statistics, for the British Association meeting in Birmingham in 1849, along with R. W. Rawson, who had earlier reviewed the *Second Report* of the Irish Railway Commissioners.

He apparently became acquainted with the English cartographer-publisher Edward Stanford, who was experimenting with the Reverend Samuel Clark on a variety of thematic maps, both atlas and wall, for the National Society for the Education of the Poor. In 1852 Petermann and Stanford had adjacent establishments in Charing Cross,[218] and Petermann did the lithographic engraving for some of the maps in the National Society's atlas.[219] In 1852 Petermann was named "Physical Geographer and Engraver on Stone to the Queen" through the influence of Baron von Bunsen, the Prussian ambassador in London. In 1854 Petermann returned to Germany to join the great geographic-cartographic establishment of Justus Perthes at Gotha, where he initiated, and for many years produced, the major periodical *Petermanns Geographische Mitteilungen*, which still ranks among the world's leading geographical journals. He remained at Gotha for the rest of his career as a cartographer-geographer occupied with the *Mitteilungen*, editions of Stieler's great *Hand Atlas*, and improving the applications of lithography to speedy mapmaking.

*Maps of People and Their Activities*

One of the thematic maps Petermann made for the National Society's atlas was a population map (FIG. 57). The map was designed by Samuel Clark, but compiled, drawn, and engraved on stone by Petermann. It is notable because the populations of every town "with more than 3,000 inhabitants is indicated by a dot the size of which is nearly as possible proportioned to the population." Whose idea this was and whether it stemmed from its being done on Harness's population map of Ireland may never be known, but this seems to have been the first instance of proportional circles being used on a thematic map that had wide distribution. The number of persons per square mile in each county is shown by a number on the map and in tables on the upper part of the sheet. Other thematic maps of the British Isles in the atlas portray the distribution of chief mineral productions, temperature, rainfall (FIG. 28), chief occupations (FIG. 67), and the ecclesiastical geography of the British Empire. The atlas was produced "For the use of Schoolmasters, Pupils, Teachers, and the Upper Classes of Schools."

Samuel Clark, M.A., F.R.G.S., is not known as a cartographer, but he should be. He was born in 1810, and at the age of thirteen and a half he had to work in his father's brush and basket business, daily from 6 A.M. to 8 P.M., after which he studied, learning French, Latin, Greek, and Hebrew.[220] He also had a full knowledge of geography. When he was twenty-six he became a partner in the publishing firm of Darton and Clark in London, and three years later, in 1839, he matriculated at Magdalen Hall, Oxford. He dissolved his partnership in 1843 and went to Italy and Greece, and because of his occupation and travel he did not receive the B.A. and M.A. degrees until 1846. Upon doing so he became vice-principal and chaplain of Saint Marks Training College, Chelsea. He was a prolific publisher. Apparently his first map publication was "Darton and Clark's School Room Map of Ireland in 1841." This was followed by a steady series of wall maps and atlases during a period of more than a quarter century, culminating in a large Bible atlas published by the Christian Knowledge Society in 1868.[221] His biographer, writing of his mapmaking, noted about his *Maps Illustrative of the Physical and Political History of the British Empire*, that:[222]

> Nothing nearly so full had ever been published before.... Clark from this time to the end of his life continued to publish a handsome series of wall-maps in conjunction with Mr Stanford and the National Society.

The British census of 1851 provided an opportunity for new population maps to be published in 1852 as part of the census report. One, a comparatively simple reference map prepared by W. Bone, shows districts and counties and includes dots (graduated circles) for all towns over 2,000 and incorporates a legend with seven dot sizes, the smallest being for towns of 2 thousand, followed by 5, 10, 25, 50, 100, 200 and ending with 300 thousand.[223] The explanation on the map states:

> Towns of Population ranging between the above numbers are indicated by dots of intermediate size proportioned to the respective populations.—The dots are not intended to represent the area covered by the Towns, but have reference entirely to Population.

Two much more sophisticated cartographic portrayals of total population to accompany the official publication of the census were prepared

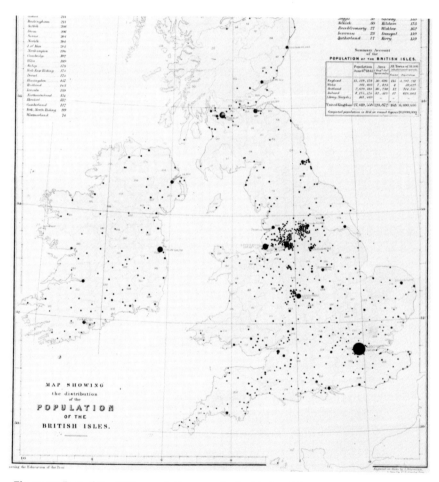

**Figure 57** Part of the 1851 population map in the National Society's atlas *Maps Illustrative*.... Reduced 50 percent. Lithograph. (Courtesy of the British Library.)

Maps of People and Their Activities

by Petermann, one of England and Wales and one of Scotland.[224] He identified himself on the map: "Designed by Augustus Petermann Phys.¹ Geographer to the Queen" and showed himself as the printer by the identification "Lith: by A. Petermann, Charing Cross, London" (FIG. 58). On these maps he combined the variable shading he had employed on his 1849 map (FIG. 56) with "black spots" for the towns as

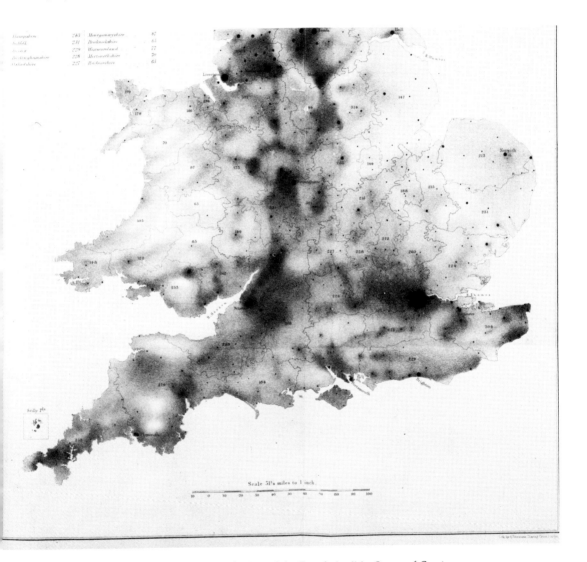

**Figure 58** Part of Petermann's 1852 "Distribution of the Population" in *Census of Great Britain, 1851*. Original scale 1:1,980,000, reduced 45 percent. Lithograph, crayon shading. (Courtesy of the British Library.)

had been done on the National Society's map (FIG. 57). Petermann's "Explanation" on the map is as follows:

> THE SHADING exhibits the various degrees of density of the population in every part of England and Wales. The very darkest shading represents a density of 600 persons and upwards to a square mile; the tints gradually becoming lighter as the density decreases—the perfectly white ground indicating a comparative absence of population. THE FIGURES denote the average amount of density of the population in each REGISTRATION COUNTY, namely the number of persons to 1 English (Statute) Square Mile. THE BLACK SPOTS represent all the towns with more than 2,000 inhabitants; the size of each spot being proportioned, approximately, to the population and the average extent of ground covered by the town.

Exactly how he decided upon the sizes of the circles is not clear from his explanation, but it does appear that he took into account both population and area as measures of "size." In any case, Petermann's two maps in the 1851 census report are highly sophisticated portrayals of very complex distributions. If we do their author the courtesy of assuming the shadings are both properly modulated and in the right places, then it would be difficult to improve the map as a display of the geographical structure of the population. Choropleth is a more commensurable form of thematic mapping but significantly less graphic. The dot map would lie somewhere between, depending upon the unit value of the dot employed.

The evolution of the thematic mapping of population seems to have progressed from the simpler systems of portrayal to increasing graphic and conceptual sophistication. Petermann's 1852 smooth shading to display variations in density is certainly more abstract than the simple choropleth map in the Irish census of 1841. It is not surprising that the most abstract thematic presentation of population density would be the last system to be devised. This is the isoline method, conceived in France but first put to use in Denmark. When used to display ratios, such as densities, the lines are technically called isopleths.

The story of the evolution of the isopleth as a method for portraying the ups and downs of a "statistical surface" is complex and is briefly outlined in chapter 7.[225] Suffice it here to note that a Danish cartographer, naval Lieutenant N. F. Ravn, prepared the first isopleth maps of population density. The two maps showed the population densities for all of Denmark for the years 1845 and 1855 and were published in 1857 (FIG. 59).[226] The maps portray densities of rural population with isopleths, but they symbolize the urban populations with circles scaled in

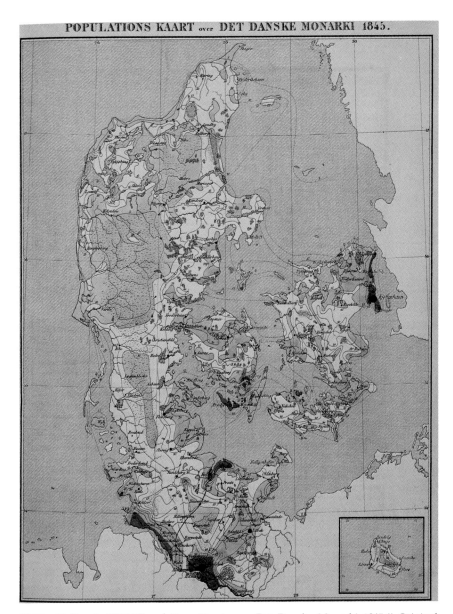

Figure 59 Ravn's 1857 "Populations Kaart over Det Danske Monarki 1845." Original 215 × 280 mm. Copperplate engraving, hand colored. (Courtesy of the Royal Library, Copenhagen.)

proportion to the populations. The specific population, that is, the density, of the number of rural persons per square geographical mile[227] was calculated for some 1,700 districts. Conceptually, a vertical with a length proportional to the density value was assumed to rise from the center of each district. The interpolated, continuous, smooth surface defined by all the end points of the verticals was then "contoured," giving the positions of the isopleths on the map.

The isopleth interval is 500 persons per square mile and the map is "layer tinted." The areas less than 1,000 per square mile are colored brown, areas from 1,000 to 2,500 are white, areas from 2,500 to 4,000 are colored light red, and those more than 4,000 are dark red. The isopleths are smoothly drawn and are continued as dashed lines across water areas. No lines are on top of one another, and no lines are left out; Ravn's maps are perfect examples of the strict application of the system.

Ravn's innovation attracted immediate attention in Germany, where Petermann in 1858 included a complimentary note about it in his "Newest Geographical Literature" in *Petermanns Mitteilungen*.[228] The term "isopleth" was suggested by Sydow the following year in a long review of the cartographic situation in Europe at that time.[229]

After the 1850s, population maps became quite common, many coming from the hand of Petermann and most of them being choropleth.[230] The first thematic mapping of the population densities of a larger area covered by unrelated censuses seems to have been by Maurice Block in 1862, in an atlas to accompany his study of the comparative power of European states (FIG. 60).[231] The map was credited as being designed by Petermann. It carries five categories of tints: less than 20, 20–30, 30–60, 60–90, and more than 90 persons per square kilometer. Cities are shown by black proportional circles.

It is remarkable that in a period of about thirty years, all the methods of portraying the distribution of population were devised and put to use. The simplest and most straightforward, the dot map, and one of the most complex, the dasymetric, both had to be reinvented.

### CHARACTERISTICS OF PEOPLES

As already noted, an interest in numbers of people was late in coming, and any cartographic expression of it had to wait for adequate census data. Curiosity about races and language came considerably earlier. In chapter 3 I referred to the four eighteenth-century maps of continents by Hensel and there reproduced his map of Europe (FIG. 15). His map of Africa (FIG. 61) shows the strong religious component characteristic of scholarly thought of that period by dividing Africa among the descendants of the three sons of Noah. The note in Latin in the lower right reads:

> All men colored yellow are indigenous peoples from the stock of Japheth. Red, descended from Shem. Green color, the possessions of Ham.

*Maps of People and Their Activities*

**Figure 60** Block's 1862 map of population density in Europe. Original 230 × 370 mm. Lithograph, printed color. (Courtesy of the British Library.)

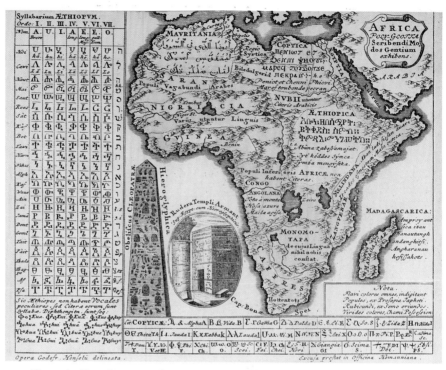

**Figure 61** Hensel's 1741 "Africa Poly-Glotta Scribendi Modos Gentium exhibens." Original 191 × 155 mm. Copperplate engraving, hand colored. (Courtesy of the British Library.)

By placing examples of the written languages of the peoples in the proper places on the map, he helped to illustrate his ideas of their origins. The tables to the left and below the map show the written forms of the sounds in Ethiopian and Coptic. Most other language and nationality maps of the eighteenth century simply had names written on them to show locations, sometimes with color added to the areas.[232]

One of the most scholarly and influential studies of language in the first part of the nineteenth century was published in 1823 by Klaproth.[233] It was accompanied by a *Sprachatlas* consisting of fifty pages of word equivalents for German and a variety of Asian tongues and one map (FIG. 62). Julius Klaproth, born in 1783, was the son of the noted German chemist Martin Klaproth, the discoverer of uranium. He early distinguished himself in language and literature, concentrating on the languages of Asia from the Caucasus to Kamchatka. His map was published in Paris, the center of intellectual life on the Continent, where he lived from 1815 until his death in 1835.

Maps of People and Their Activities

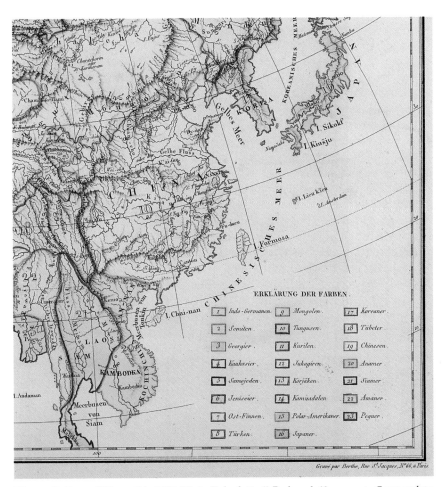

Figure 62 Part of Klaproth's 1823 "Asia Polyglotta." Reduced 40 percent. Copperplate engraving, hand colored. (Courtesy of the British Library.)

Religion was a subject charged with an unusual amount of emotion in the period when thematic cartography was evolving. The various reform movements and the relations among civil authorities and religious adherents made the topic always volatile. It is somewhat surprising, therefore, that maps portraying the distributions of faiths and denominations in an area as complex religiously as western Europe seem not to have been very common. Eckert mentions a "Carte ecclésiastique de l'Empire Français..." made in 1808,[234] and Wyld's 1815 "Chart of the World Shewing the Religion, Population and Civilization of Each Country" was earlier referred to as one of the first maps to show population figures. In Britain the Catholics were emancipated in 1829 and in

1833 the British Reformation society published a kind of dot map entitled "A Map Showing the Situation of Each Roman Catholic Chapel, College, or Seminary throughout England, Scotland, and Wales according to Published Accounts and Also the Present Stations of the Reformation Society to January 1833." Three separate symbols were used, one for each item in place. There being too many crowded together in Lancashire, that area was enlarged as an inset.

Germany was, of course, the most complex area, and this is reflected in a rather complicated thematic map by J. V. Kutscheit published in Berlin in 1845 (FIG. 63). Colored bands along the boundaries help to identify sixteen states. A set of seven intermediate ratios, from almost completely Catholic to almost completely Protestant, are symbolized by intricate black and white patterns, along with more than twenty point symbols (FIG. 64). This is among the earliest maps, if not the first, to portray a range of ratios between the components of totals.

Both Berghaus's *Physikalischer Atlas* and Johnston's *The Physical Atlas* included world maps of religions and by the 1860s the mapping of religions had become standard. Some, like Kutscheit's map, were rather complicated. For example, in 1860 A. Hume prepared a choropleth map showing by parallel line and cross hatched patterns the proportions of religious to total populations.[235] The cities were represented by circles

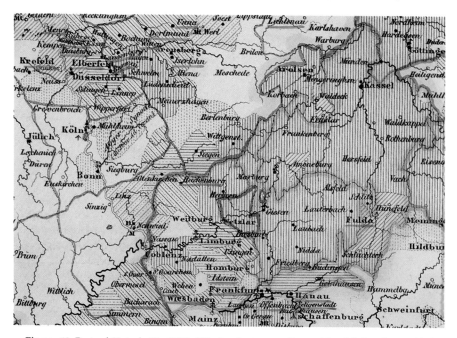

**Figure 63** Part of Kutscheit's 1845 "Kirchenkarte von Deutschland." See figure 64 for legend. Nearly full size. Lithograph, hand colored. (Courtesy of the British Library.)

Maps of People and Their Activities

(all the same size) which were subdivided with radial lines to show percentages of (1) churchmen, (2) Protestant dissenters, (3) irreligious or nominal churchmen, and (4) Roman Catholics. Block's atlas of 1862, referred to above, included a map of the religions of Europe.

Some subjects are prone to dogmatic treatment, and certainly religion is one such. For example, Berghaus included a section on anthropog-

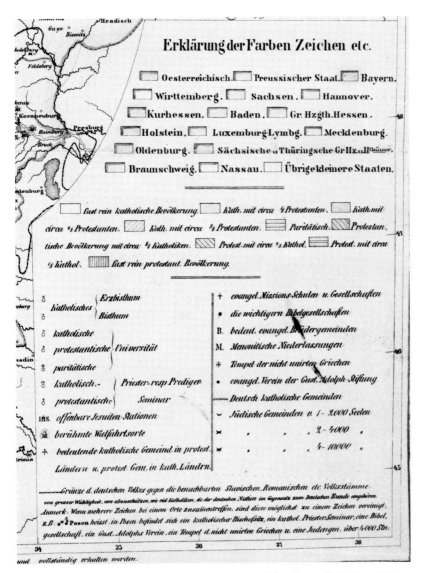

**Figure 64** Legend for Kutscheit's "Kirchenkarte...." See figure 63. Reduced 25 percent. Lithograph, hand colored. (Courtesy of the British Library.)

raphy in the second major section of his *Physikalischer Atlas* published in 1848. One of the plates in that section included four separate maps showing various aspects of anthropography: forms of occupation, religions, forms of government, and spiritual constitution. The fourth is shown here as FIGURE 65, and it has an interesting legend which in translation reads as follows:

> SPIRITUAL CONSTITUTION
> The maximum is in the Christian lands, preferably of the Protestants; therefore these are shown as light (white). The minimum or complete darkness of the soul is in the territory of the heathens, shown here as darkest, with the transition indicated by shading.

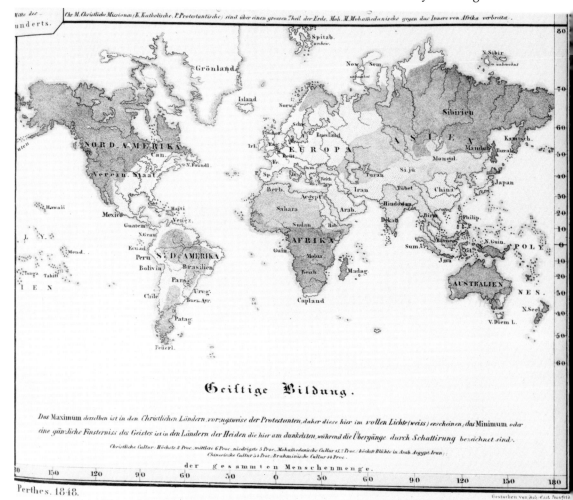

**Figure 65** World map of "spiritual constitution" in Berghaus's *Physikalischer Atlas*, seventh Abteilung, Anthropography, no. 4. Reduced 22 percent. Copperplate engraving. (Courtesy of the British Library.)

## Maps of People and Their Activities

A note at the bottom gives the percentages of the total population of the earth having Christian, Mohammedan, Chinese, and Brahmin cultures.

Curiosity about the distribution of peoples is probably quite old, and its expression on maps in the form of pictorial representations goes back a long way. In more modern times, such as the mid-seventeenth century, the drawings became somewhat more instructive rather than mere embellishments, as for example in some of Blaeu's maps in his *Atlas Novus* (1635), where they attempt to portray anthropological characteristics.[236] The curiosity, however, did not, apparently, result in systematic mapping until the nineteenth century, if we discount the biblical divisions of the earth among the three sons of Noah which began in the earliest figurative maps of the so-called Dark Ages and even survived to a degree on Hensel's 1741 language map of Africa (FIG. 61). One would expect that the mapping of the distribution of varieties of the human species would have followed naturally soon after Linnaeus's *Systema Naturae* (1735), Buffon's *Histoire Naturelle* (1749 ff.), and their extension by the German anthropologist Blumenthal and even the French naturalist Cuvier in the early nineteenth century. Nevertheless, it did not.

Few subjects seem to be so intricate or to have been contemplated from so many naive points of view, especially in the early period when distinctions among skin color, geographical location, languages, and a host of other characteristics tended to be rather haphazardly mixed in trying to classify *Homo sapiens*. Ritter's 1806 pedagogical atlas, *Sechs Karten von Europa*, includes one map (map 6), which, in addition to population numbers and densities plotted here and there, contains names of peoples. These names are a mixture of nationalities, languages, and ethnic groups, and their regions of occurrence are only implied by where the names are written on the map. The earliest anthropological map referred to by Eckert is one by C. F. Weiland which appeared in 1835.[237] Another of the physicians who tried his hand at thematic mapping was the ethnologist J. C. Pritchard, who published in 1843 a set of ethnographical maps to accompany his classic work *The Natural History of Man*.[238]

The most widely distributed ethnographic map at that time was the work of Gustav Kombst, a German living in Edinburgh. Kombst, on good relations with the Johnston's publishing house, began acting as a go-between for them and for Berghaus in 1841 regarding the issuance of an English edition of Berghaus's *Grundriss der Geographie*... and some of Berghaus's physical maps.[239] Kombst's ethnographic map of Europe (FIG. 66) appeared in Johnston's *National Atlas*... in 1843 and was described by its author as:[240]

> the first Ethnographic Map ever published, based
> upon the principle of the natural physical difference of
> the different varieties of the Caucasian species in-
> habiting Europe.

It also was included in Berghaus's *Physikalischer Atlas* and Johnston's *The Physical Atlas*.[241] The eighth Abteilung, Ethnography, of the *Physikalischer Atlas* contains nineteen maps covering many parts of the earth in comparative detail and indicates the rapidity with which that

**Figure 66** Part of Kombst's 1843 "Ethnographic Map of Europe" in Johnston's *The National Atlas*.... Reduced 50 percent. Copperplate engraving, hand colored. (Courtesy of the British Library.)

*Maps of People and Their Activities*

class of thematic maps almost burst upon the scene during the 1840s.

Kombst's ethnographic map of Europe is interesting cartographically in the systematic and logical use of color. In a note to the map in Johnston's *National Atlas*... the use of color is described as follows:[242]

> The three great varieties of the Caucasian species have been pointed out, the *Celtic* by *blue*, the *Teutonic* by *yellow*, and the *Sclavonian* by *red*. The subvarieties of these varieties have different shades of these fundamental colours. Wherever there has been a crossing of these varieties or subvarieties, it is indicated by a *mixed* colour, in such a manner that the colour predominant in the mixture points out the predominant national element. Thus green, in its different shades, points out a mixture of Celtic and Teutonic blood; flesh colour, and other tints mixed of red and yellow, etc., point out a mixture of Teutonic and Sclavonian blood. The same principle applies also to the varieties of the Mongolian species which inhabit Europe.

Kombst and Johnston also faced squarely the perennial cartographic problem of mapping the geographical occurrences of intermixtures and transitions in a classification system, by observing:[243]

> So far as the colouring of the map is concerned, it may be objected, that the different populations cannot always be separated from each other by distinctly drawn lines, because, as is mostly the case in nature, they blend so much into each other. This the author is fully aware of. But he preferred giving a *distinct outline*—even if it should be objectionable for the reason stated—to *an overstudied accuracy in the colouring*, which gives only a confused representation of the distribution of the population, and holds out to the spectator's eye an image, which, however much it may be appreciated by a profound connoisseur of the subject, necessarily bewilders and leads astray those whose object it is to acquire knowledge.

The intellectual boldness of the period is illustrated by Kombst's inclusion of summaries of the intellectual and moral character as well as the physiological descriptions of his varieties of the Caucasian species. One can admire his temerity while doubting his judgment. They make for good reading, and unfortunately there is space for only one here. His description of the intellectual and moral character of the Celtic variety is as follows:[244]

> Quickness of perception, great powers of combination and application, disposition for concentrating power,

love of equality, of society, of amusement, of glory, want of caution and providence, prevalent disposition for sexual intercourse, gallantry, want of respect for human life. Want of penetration, of desire for personal independence and political and civil liberty; national pride. Disposition to superstition and hierarchy. Foundling hospitals. In most Celtic countries no law regarding paternity or for the sustenance of so called natural children by the father. Fine blandishing manners, great external politeness, without inward sympathy. Irascible, not forgetful of injuries, little disposition for hard work. Bad seamen, and not fit for colonizing.

Descriptions of the Teutonic and Sclavonian varieties are equally blunt and dogmatic. It was probably a good thing for his own personal safety that Kombst devoted much attention to showing mixture on his map.

### Economic Activities

In general, maps of occupations, production, and trade were among the last of the maps to appear during the developmental period of thematic cartography. This tardiness was probably largely due to a lack of statistical data. Such information was not collected in a central place, in most instances, simply because of a lack of official interest. The concept of a national economy worth investigating from a geographical point of view had to await the gearing up of industrial production, the trade generated by surpluses, and the social and political stability that would allow these to develop. Such conditions began to coincide in western Europe only after the industrial revolution had gathered momentum, after the turmoil of the French Revolution and the Napoleonic wars, and after the era of road, canal, and railway building had gotten well under way. Accordingly, we would not expect to find much in the way of thematic maps of economic activity before the second quarter of the nineteenth century.

Students of early thematic cartography have noted a few such maps produced in the eighteenth century, but these are generally rather primitive thematic maps which show locations of individual phenomena or overall structural relationships among areas and activities. Often these maps are manuscripts which, fortunately, have been preserved in map collections.[245] One of the more interesting of the early maps is one described by Fick which was published about 1730.[246] Its author was Matthäus Seutter, who learned to be a cartographer in the establishment of Homann in Nürnberg but in 1707 formed his own business in Augsburg. There he, and his family after him, produced

*Maps of People and Their Activities*

many maps and atlases. Among the mostly general maps produced by Seutter is one prepared as an instructional device to help teach map reading. It shows an imaginary region where a kind of ideal landscape has developed on which to show the important geographic phenomena, a large share of which are economic. Of the forty symbols, eighteen refer to economic items such as various kinds of mills, mines, and similar classes of activities. A great many of the general maps of actual areas made during the eighteenth century included a variety of symbols to show such items as appeared on Seutter's instructional map, but they can hardly be classed as thematic maps, for their primary function was to show the locations of administrative boundaries, roads, and place names. As time went on, the symbols multiplied until they became the primary features.

The best known of these maps, and often called the first economic and thematic map, is Crome's "Neue Carte von Europa..." (FIG. 16) referred to in chapter 3.[247] It shows the locations of more than four dozen products. It was greatly outdone by the "Natur und Kunst Producten Atlas der Oestreichischen, Deutschen Staaten," published in Wien in 1796 by H. Blum, which consisted of a series of maps and cities, each with a variety of symbols.[248] The "Natur und Producten Karte von Oestreich ob der Ems" has a legend with twenty-eight signs showing cities and associated features, sixty symbols showing the natural products of mines, forests, and lakes, and thirty-eight signs showing the locations of economic activities such as brewing and charcoal-making. This great variety of marks was put on a map which also showed terrain with a primitive kind of hachuring, all done in black and white (the *Kreise* were colored). Needless to say, the map is almost illegible.

These kinds of product-location maps continued to be made during the first four decades of the nineteenth century. Except for two very unusual flow maps by Harness, published in 1838, thematic mapping of economic activity appears to have lagged well behind other subject areas until the 1840s, when it suddenly burgeoned.[249] There are only a few references to such maps in the early 1840s. One is to an industrial map of Bohemia produced in Prague by E. von Schwartzer in 1842,[250] and another is the "Strassen-, Industrie, und Verkehrskarte von Zentral-Europa," Berlin, 1846 by J. V. Kutscheit, based on the first Berlin Industrial Exhibition in 1844.[251] The first map by Petermann to appear under his own name also was occasioned by the Berlin Exhibition.[252] It is an economic-geographic map of Germany made by Petermann in 1844 while he was still Berghaus's student, before he emigrated to Scotland. In general these maps used symbols to show the various industries, though Petermann did use area coloring to show the Wupper industrial region on an inset map. Two years before the Berlin Exhibition, a somewhat more complex economic-geographic map of

European Russia had appeared in Saint Petersburg, and a German edition was published in 1844.[253] The Russian map used colors as area symbols to show forested, industrial, agricultural, and grazing areas, along with circles containing letters to show the kinds of industrial production. Darker colors in the circles set them off from the background colors.

Far more sophisticated maps of economic geography began to appear in the early 1850s. One of the first is the map of chief occupations of the peoples in the British Isles (FIG. 67). This map, included in the National Society's *Atlas*, was designed by Samuel Clark and compiled and drawn by Petermann.[254] It portrays the agricultural, pastoral, and manufacturing regions with colors and patterns and even distinguishes five classes of manufacturing areas. It shows by symbols a variety of products and the populations of four categories of cities. The map also displays such information as the northern limit of wheat, the principal salmon fisheries, and the primary activity of the chief manufacturing centers. The Clark-Petermann map is a notable improvement over the earlier economic maps, but it is as much a general map as it is thematic, since it shows such a variety of information.

The year the Clark-Petermann map appeared, 1851, was also the year when the first of the great international expositions was held in London at the famed Crystal Palace. That it could occur at all, that more than 13,000 exhibitors came from many parts of the world, and that it attracted some 6,000,000 visitors is impressive evidence of the enormous changes that had taken place in the previous few decades in manufacturing, transportation, and trade. Such an affair twenty-five years earlier would have been unthinkable: there would have been little to exhibit, and for anyone to travel any great distance, such as from central Europe, would have required a long and tedious journey.

Petermann, having moved to London and being established then in his own mapmaking business, made a relatively large panel of eighteen small maps entitled "Geographical View of the Great Exhibition of 1851 Shewing at One View the Relative & Territorial Distribution of the Various Localities from Whence the Raw Materials & Manufacturers Contributed to the Exhibition Have Been Severally Supplied."[255] The general regions from which contributions came to the exposition are designated by black lining, but specific specialized manufacturing areas are portrayed with watercolor inside fine bounding lines. Raw materials and manufacturers are shown with a variety of individual point symbols devised in such a way that the assemblage at any place might be easily combined.[256]

A sign of growing maturity, of sorts, in a field is the increasing tendency to specialize. In the thematic mapping of economic activities this is revealed in the trend toward portraying fewer phenomena on the

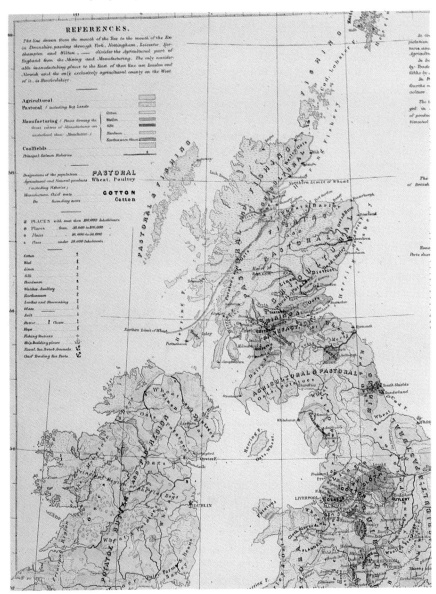

Figure 67 Part of the Clark-Petermann 1851 map of chief occupations in the National Society's atlas *Maps Illustrative* .... Reduced 35 percent. Lithograph, hand colored. (Courtesy of the British Library.)

map. The maps of the 1840s and early 1850s, with their manufacturing regions, were considerably more sophisticated than the economic-geographic maps of former times which depended largely on a multitude of individual symbols. However carefully done, such earlier presentations were little more than general reference maps, since there was little portrayal of the organization of regional variations of economic activity. Cartographic portrayal of regionalization and concentration on individual activities or commodities began in the late 1840s and by the mid-1850s data were good enough that more specialized maps could be made.

These kinds of economic maps were uncommon, and they were not widely circulated. In previous sections I have noted that the physical and natural scientists were generally well acquainted with one another, through correspondence, membership in societies, and the exchange of information. So, too, were the engineers. The curriculum at Woolwich required French, and we have noted Baron Charles Dupin's lengthy study of Britain's power base. Guerry's studies in France and Quetelet's investigations of the social physics of man were well known to all the statisticians of the 1830s and 1840s, and they made a considerable impression on Larcom in Ireland. Quetelet's major work was translated from the French into English and German within seven years of its appearance in 1835. The geographers, with the partial exception of Humboldt, Berghaus, Petermann, and some of the other German scholars, were largely concerned with exploration and physical science. Clearly, in the mid-nineteenth century natural history (including man), geographical exploration, and the geophysical sciences dominated scholarly curiosity and, consequently, the cartographic portrayals of that interest.

Nothing could better illustrate this bias than the life and works of a French engineer turned economic geographer-cartographer. Space does not allow a detailed description of the life and maps of Charles Joseph Minard, an ingenious and prolific cartographer whose works seem to have been well known to statisticians and students of economic affairs in France, but not to geographers.[257] Minard was born in 1781 and entered the renowned Ecole Polytechnique when he was only sixteen, from which he transferred to the Ecole Nationale des Ponts et Chaussées, the training school for engineers responsible for building and maintaining the port facilities, roads, canals, and later the railways of France. He rose through engineering activities in many parts of France to become superintendent of the Ecole Nationale des Ponts et Chaussées and finally to the rank of inspecteur général. In the 1830s Minard became increasingly interested in economic geography, especially that having to do with the national and international movements of goods. From 1845 until his death in 1870 he produced some fifty maps, the majority of which are flow maps, to which we will turn later.[258] Minard made a few nonflow economic maps, two of which I show here to illustrate his relatively advanced technique.

In 1858 Minard prepared a map to show the amounts of butcher's meat supplied by each *département* to the Paris market (FIG. 68). The

*Maps of People and Their Activities* 145

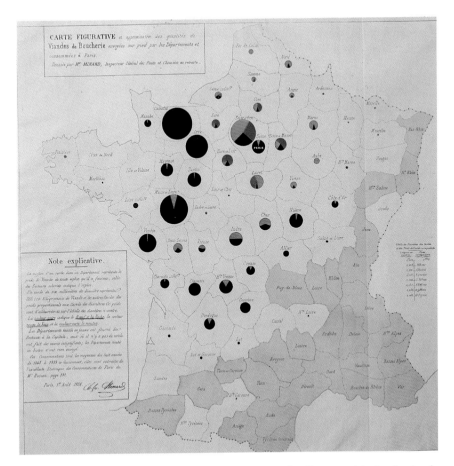

**Figure 68** Minard's 1858 "Carte figurative et approximative des quantités de viandes de boucherie envoyées sur pied par les départements et consommées à Paris." Original 530 × 520 mm. Lithograph, hand colored. (Courtesy of l'Ecole Nationale des Ponts et Chaussées, Paris.)

contribution of all kinds of meat from each *département* is shown by a circle, and the circles are scaled precisely so that their areas are in direct proportion to the weighted quantities of meat they represent. The circles are subdivided to show the relative amounts of beef (black), veal (red), and mutton (green). The *départements* tinted yellow supplied meat to Paris, those that are yellow but without a circle provided only insignificant amounts, and those tinted bister supplied none. Minard stated that the data mapped were the means of the eight years from 1845 to 1853 inclusive.[259] In 1861 Minard made a map showing the market areas for foreign coal and coke in France in 1858 (FIG. 69). Using five tints, he showed the areas supplied by England (green), Belgium (blue),

Germany (tan), England and Belgium (dark green), and Belgium and Germany (dark blue). The colors are well chosen to symbolize single and double sources.

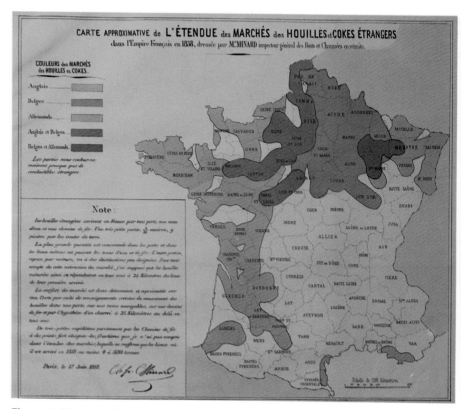

**Figure 69** Minard's 1861 map of the extent of the markets for foreign coal and coke in France in 1858. Original 418 × 361 mm. Lithograph, hand colored. (Courtesy of L'Ecole Nationale des Ponts et Chaussées, Paris.)

By the early 1860s sufficient data were available and the printing techniques had developed to the point where an economic atlas could be produced, such as Block's atlas to accompany his study of the comparative power of the several states of Europe.[260] The national areas are the enumeration districts for the choropleth maps. Map 1 in the *Atlas* (FIG. 60) and four others, map 2—Population growth, map 3—Sizes of armies, map 4—Sizes of navies, and map 12—Religions, deal with non-economic elements, but the other seven are choropleth, thematic maps on the following subjects:

Map 5. Total national incomes
6. Average taxes paid by each inhabitant
7. Sizes of the public debt per inhabitant

8. Interest rates on the public debt
9. Relations between lengths of railways and the areas of countries
10. Values of foreign trade per inhabitant
11. Average rates of customs duties.

These maps were all printed in color, and each distribution is divided into classes and shown with tints of a color. For example, map 9, portraying the relations between the lengths of railways and the areas of countries, employs five classes: states having less than 10 km of railway per 1,000 square km, 10 to 20 km, 20 to 30 km, 30 to 40 km, and more than 40 km (FIG. 70). The map clearly shows the commanding lead in rail transport that Britain enjoyed, even at that late date. The maps were produced at Gotha under the design supervision of Petermann. In addition to national boundaries, the populations of cities are designated by graduated circles and categorized in five classes by sizes and styles of lettering. Railways, navigable rivers, and canals are included in the base data.

MOVEMENTS OF GOODS AND PEOPLE

As we have traced the development of each subject matter or class, or each functional category of thematic maps, it is apparent that most have followed a kind of normal growth process. The tendency has been for the first ones to be rather simple and tentative, and for later ones to progress to a stage not greatly different from what is done today. Initial growth in sophistication was often slow, but the rate of development accelerated, since later innovations could borrow and adapt techniques. One class of maps does not fit that generalization in the least—the flow map.

The first published flow maps, so far as is known now, were made in 1837 by Harness, who prepared a population map (FIG. 53) and two flow maps for the *Atlas to Accompany the Second Report of the Railway Commissioners, Ireland*.[261] One of the flow maps portrayed the average number of passengers carried in one direction weekly by public conveyances by means of shaded lines, the widths of which Harness clearly intended to be directly proportional to the numbers of passengers.[262] The numbers appear alongside the lines, and the entire map has a most modern appearance. Harness's other flow map was of similar design "Showing the relative Quantities of Traffic in Different Directions" (FIG. 71). One reviewer of the *Second Report of the Railway Commissioners, Ireland*, did mention the maps,[263] but Harness's flow maps seem then to have dropped from sight and remained essentially unknown for nearly a century. Of this we cannot be sure, however, because in the mid-1840s Alphonse Belpaire in Belgium and Minard in France began making flow maps at about the same time. Both were engineers concerned with railways and may well have seen or have been told of the Harness maps.

**Figure 70** Block's 1862 map of the relation between the lengths of railways and the areas of countries in Europe. Original 230 × 370 mm. Lithograph, printed color. (Courtesy of the British Library.)

Maps of People and Their Activities

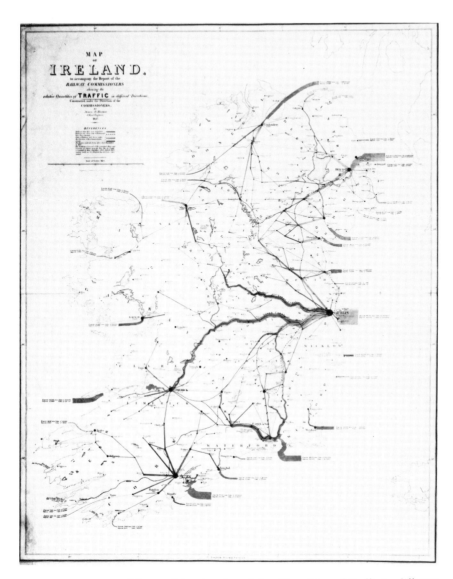

**Figure 71** Harness's 1837 map "showing the relative Quantities of Traffic in different Directions." Original 1:633,600. Copperplate engraving. (Author's collection.)

Belpaire was an engineer of the Ponts et Chaussées in Belgium who was born in 1817 and lived only to age thirty-seven.[264] In the mid-1840s he made three very large flow maps of Belgium. Two of these maps, one each for movement in 1834 and 1844 showed the amount of traffic by bands or flow lines scaled in "units of transport," each of which was allocated a width of 0.5 mm.[265] One unit of transport was 10,000 tons of bulk merchandise, 5,000 tons of baggage or packaged materials, or

30,000 passengers. Blue was used for roads, green for rivers and canals (with gray for the coal component), and red and yellow for railways—red for passengers and yellow for merchandise. Numbers along the bands noted the numbers of units.

Neither of the maps bears a date of preparation or publication, but among Belpaire's publications is a monograph, dated 1847, concerning his maps of the traffic in Belgium, and in this he points out that he published the maps "about two years ago."[266] He also made a similar map of movement on land routes during the year 1843,[267] of which a section showing the region of Bruxelles, Malines, and Louvain was later published with the title "Extrait de la carte du mouvement des transports en Belgique, publiées en 1843 par M. l'Ingénieur Belpaire." If one assumes the title of this map to be correct, then Belpaire's first flow map would have been published in 1843. On the other hand, it could not reasonably have been published the same year as the date of the data. Even if that had somehow been possible, one would expect Belpaire to have been more specific about it in his monograph. We shall probably never know the exact publication date, but there is no question that Belpaire in Belgium and Minard in France produced flow maps at about the same time some seven or eight years after Harness's map had appeared.

The question of priority aside, Minard clearly outdid Harness and Belpaire in the number, variety, and sophistication of his thematic maps of movement.[268] Of the fifty-one thematic maps Minard produced, forty-two are flow maps. More than a dozen make up a general series of very detailed, complex, reference maps of the tonnages of merchandise which circulated annually in France on all major routes of transport.[269] Minard's flow maps covered a wide range of subjects: people, armies, cereals, wines, spirits, livestock, coal, coke, and fibers, as well as general merchandise. They were all lithographs, and the earlier ones were hand colored. Later a large proportion employed printed color. Space does not allow reproduction of enough of Minard's flow maps to characterize effectively the graphic ingenuity of Minard the thematic cartographer. They range from many maps of France and the world to a few of Europe. One each of these categories must suffice.

Minard made a series of maps showing worldwide movements of merchandise, such as coal exported from Britain, cotton importations into Europe, and even the countries of origin and destination of immigrants. One such world map showed the export of French wines by sea in 1864 (FIG. 72). It illustrates an interesting characteristic of Minard's maps, namely that he was much more concerned with portraying the basic structure of the distribution than he was with maintaining strict positional accuracy of the geographical base—this from an engineer!

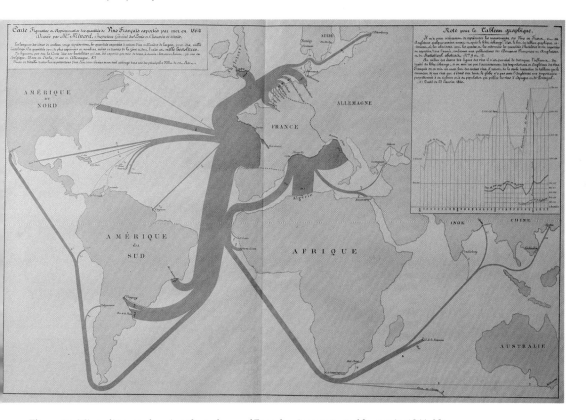

**Figure 72** Minard's map showing the volume of French wines exported by sea in 1864. No date. Original 835 × 547 mm. Lithograph, printed color. (Courtesy of l'Ecole Nationale des Ponts et Chaussées, Paris.)

The logic is unassailable: if a flow line of a given width is to portray ocean transport it would be as ridiculous to have it cross land, such as the margins of the Channel or the Straits of Gibraltar, as it would be to plot a railway across water to avoid overcrowding. The solution was simply to widen the Channel or the Strait. Minard always included the terms *carte figurative et approximative* whenever it was appropriate, which was in well over 90 percent of his fifty-one maps.

One of his maps of France displays the weight of livestock carried to Paris by railway in 1862 (FIG. 73). The main map employs four colors to distinguish among cattle *(boeufs et vaches)* in yellow, calves in red, swine in gray, and sheep in blue. The weights are indicated by the widths of the colored lines in the ratio of 1 mm to 1,000 tons. The weights were equated as cattle—300 kg per head, calves at 70, swine at 100, sheep at 20. The map was made to illustrate the effects of railways on the traffic. The note in the upper right discusses the movements in

1828, before railways were built, in relation to the movements in 1862. Not only was the movement greatly augmented, but the per capita consumption rose as well. The small inset map shows the changes in the extent of the areas from which cattle *(boeufs)* were dispatched to Paris in 1828 and 1862, the pink denoting the extension by the latter year.

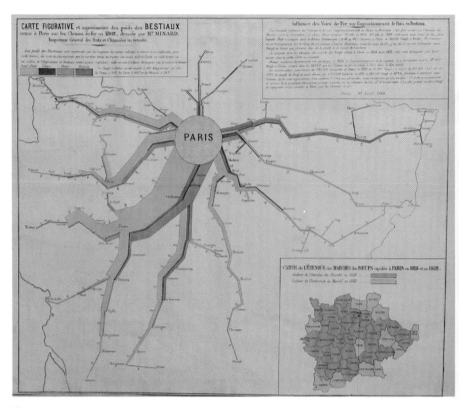

**Figure 73** Minard's 1864 map showing the weight of livestock that came to Paris by railways in 1862. Original 555 × 470 mm. Lithograph, hand colored. (Courtesy of l'Ecole Nationale des Ponts et Chaussées, Paris.)

One of the more interesting maps made by Minard portrays passenger traffic on the principal railways of Europe (FIG. 74). The amount of traffic is indicated by the widths of the flow lines in the ratio of 2 mm per 100,000 travelers and is also noted on the map by numbers written at right angles to the lines, unity being 1,000 passengers. A legend-table at the left shows each state by color and lists for each its total area in square kilometers, its total length of railways in kilometers, and, in column 3, the number of meters of railway per square kilometer. In an explanation of the map he states his philosophy of *cartes figuratives*:

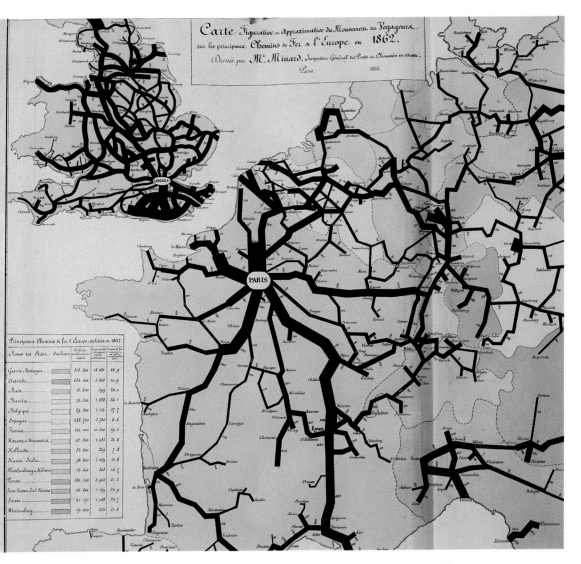

**Figure 74** Part of Minard's 1865 map of the movement of travelers in 1862 on the principal railways of Europe. Reduced 80 percent. Lithograph, hand colored. (Courtesy of l'Ecole Nationale des Ponts et Chaussées, Paris.)

In order to position the flow lines without enlarging the map too much, which would hinder encompassing it at a glance, which is the advantage of *cartes figuratives*, I have had to alter the geographical proportions considerably and to omit Ireland and Scotland.

Minard's maps were widely used and his system copied in France. His first map, published in 1845, showed the movement of passengers by public carrier on the route where a railway was being considered between Dijon and Mulhouse, and was distributed in two hundred copies. He reports that, soon after, other engineers imitated it to support their projects.[270] He also pointed out in the same pamphlet that he had published nearly ten thousand copies of his graphs and *cartes figuratives* and that even the emperor was impressed by them. Although Minard's thematic maps were noted by such writers on statistics and graphical method as Block and Marey as well as by the International Statistical Congress meeting at Vienna in 1857,[271] few others seem to have noted them, and Minard, like Harness and Belpaire, was soon forgotten. Like the dot map and the dasymetric technique, their sophisticated cartographic methods would have to be reinvented.

# *Six*

# MAPS OF THE

# SOCIAL ENVIRONMENT

THE TIME OF RAPID DEVELOPMENT IN THEMATIC CARTOGraphy began in the latter part of the 1600s and extended well into the 1800s. The period was far from placid. Social unrest had been growing for some time and finally erupted in the upheaval of the French Revolution. The urge for social justice caught the fancy of many idealists, such as Humboldt and his friend Georg Forster, but the seeds of reform were sprouting almost everywhere. Part of this was the result of a growing idealism, but much of it stemmed from changes as a consequence of the spreading industrial revolution. Movements to the cities created obvious problems of food, housing, and health that could no longer be ignored. The Napoleonic wars and the growth of trade turned attention to the state as a living institution, and studies in political economy probing the bases of national power were more and more common.

As we noted before, the early part of the nineteenth century marks the first time social statistics began to be gathered on a large scale, and as these numerical data became available the field of statistics as a separate technique and body of thought began to emerge. The manipulation of numbers was eagerly undertaken by people trained in many fields. Doctors, lawyers, engineers, scientists, and self-made scholars of all kinds formed local statistical societies, many of which published journals whose contents attest to a growing fascination with numbers of almost any kind. When we survey this period in the first half of the nineteenth century, one notable scholarly characteristic is that what we now call cultural and economic geography was being practiced by all sorts of people, but not much by those called geographers at that time. The geographers were the explorers and the official mapmakers, but it was mostly others who were making thematic maps of various distributions with the hope of discovering meaningful correlations among the regional variations. The interrelation of geographical distributions had become an increasing focus of curiosity since it had first begun to be systematized by Humboldt and Ritter. With the availability of numerous social statistics, the intriguing study of correlations was no longer restricted to the natural scientist, investigating, for example, the relation between differences in temperature and flora; instead, one could look into such interesting things as the relation between variations in crime rates and education, a favorite subject in the category that came to be called moral statistics.

In the first half of the nineteenth century Europe was visited by a series of cholera epidemics. Mortality rates were high, and, since the cause of cholera was not known, interest tended to focus on the more obvious possibilities, the environmental conditions under which the majority of the afflicted lived. Medical mapping in Europe burgeoned in the 1830s when cholera first appeared.[272] It became increasingly common thereafter, and the use of thematic maps as tools both to suggest and to test hypotheses became quite sophisticated in the hands of the medical investigators. With or without cholera, the appalling living conditions of the poor, who made up a very large segment of the population, could not escape notice, and various aspects of housing, land use, and population characteristics began to be mapped.

We shall begin our review of the development of thematic mapping of the social environment with the earliest category, moral statistics.

### Moral Statistics

The designation "moral statistics" appeared for the first time in 1833 in the title of the book by André-Michel Guerry: *Essai sur la statistique morale de la France*, but maps on the subject had appeared earlier. The term encompasses a wide array of characteristics of populations, from statistics on education and crime, to pauperism, and even to "improvident marriages," a reproachful term widely used in the nineteenth century for marriages entered into by males under the age of twenty-one.

The first map of moral statistics, "Carte figurative de l'instruction populaire de la France," was made in 1826 and published in 1827 by Charles Dupin (FIG. 75).[273] Dupin, born in 1784, entered the Ecole Polytechnique in 1801, and joined the Génie Maritime, in which he ultimately became an inspector-general. An accomplished student of geometry, he gained admission to the Académie des Sciences in 1818 and became a professor (part time) in the Conservatoire des Arts et Métiers in 1819. He became increasingly interested in political economy and spent three years in Great Britain studying the bases of its commercial power.[274] As time went on he devoted more and more time to politics, being elected to the Chamber of Deputies in 1828 and named a senator in 1853.[275]

Maps of the Social Environment

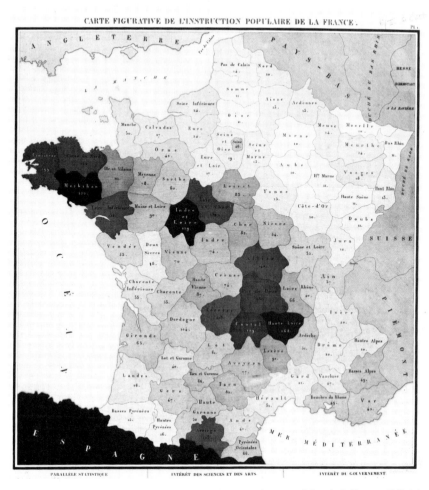

**Figure 75** Dupin's 1827 "Carte figurative de l'instruction populaire de la France." Original ca. 510 × 535 mm. Lithographic engraving. (Photo. Bibliothèque Nationale, Paris.)

On 30 November 1826 Dupin delivered an address to the Conservatoire des Arts et Métiers on the general subject of popular education and its relation to the prosperity of France. On that occasion he exhibited his map, which appears to have initiated the cartographic portrayal of moral statistics.[276] It also is the first now-known choropleth map, that being the name generally applied to statistical maps in which the enumerated data are assumed to be averages applying equally to all sections of the appropriate enumeration districts. Some sort of graphic distinction among the districts, such as shading or color, portrays the variation among the averages.

Dupin attacked the problem of equating the statistics, referred to earlier, in a curious and unique fashion, the numbers inscribed in each *département,* and the basis for the shading, being the number of persons in the *département* per male child at school. The map is interesting for technical reasons, which will be discussed in chapter 7, and its graphic portrayal apparently had great impact. His biographer in *La Grande Encyclopédie...* wrote:

> His *Carte de la France éclairée et de la France obscure,* in which he was the first to have the idea of showing by tints the proportion of the illiterates in each *département,* brought him deserved fame from then on.

It was referred to repeatedly and a remarkably similar map was prepared by Somerhausen for the Pays Bas using data of 1827 (FIG. 20).[277] In 1864 Guerry, to whom I will turn shortly, noted that Dupin's work more than anything else excited attention and contributed strongly to the increase in the number of schools in France.[278]

Dupin made other thematic maps, but none so influential as his 1827 map of popular instruction. In the decade following the appearance of that map both Guerry and Quetelet made maps of moral statistics having to do with crime, and this may have prompted Dupin to try his hand at it. At any rate, he attended the meeting of the Statistical Section of the British Association for the Advancement of Science meeting at Bristol in 1836, presented a paper dealing with the effect of the price of grain on the population of France, and later, "exhibited a map of England, illustrating the proportion of crime to the density of population."[279] The society elected Dupin a foreign honorary member in 1837–38. In 1843 he prepared another *carte figurative* (accompanied by a short paper) showing for each *département* the relation among the area in vineyards, the area planted with sugar beets, and the total area.

In 1829 the first maps of crime were made by Balbi and Guerry from the criminal statistics of the Compte Général de l'Administration de la Justice Criminelle for the years 1825, 1826, and 1827 and the data from the latest census.[280] These appeared as a simple display of three maps: crimes against property, crimes against the person, and instruction (FIG. 76). Like Dupin, the authors equated the statistics for areal variation by expressing them in terms of the number of persons per crime for the court divisions or per scholar for the educational districts.

Balbi was a prolific writer of "geography," being much concerned with ethnography and languages, but his one collaboration with Guerry seems to have produced no lasting effects. All his other maps were general. Guerry was very different.[281] He was born in 1802 and studied law, becoming *avocat* at the law court in Paris, where he became inter-

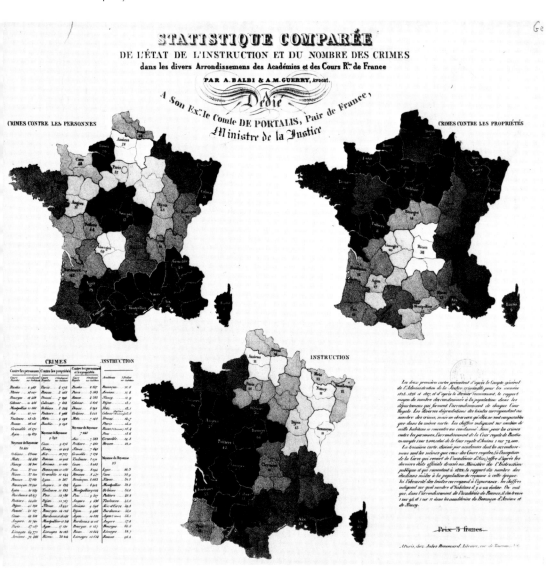

Figure 76 Maps by Balbi and Guerry (1829) comparing statistics regarding instruction and incidence of crimes. Reduced 70 percent. Lithograph, crayon shading. (Photo. Bibliothèque Nationale, Paris.)

ested in criminal statistics and the causes of crime.[282] Guerry was to become increasingly active and well known for his studies and maps regarding moral statistics, and I will return to him later, but he was preceded in his next study by the Belgian Adolphe Quetelet.

Quetelet was a powerful intellectual figure in the second and third quarters of the nineteenth century. He was a leader in statistical and sociological theory, he corresponded with royalty, and his written words had a potent influence on man's thoughts about himself. He was born in 1796 at Ghent. He studied mathematics, in which he became proficient enough to teach it at a college in Ghent when he was only nineteen. He studied astronomy at the Observatory at Paris, and the theory of probability under Laplace. He directed the Royal Observatory at Brussels and became the leading statistician in western Europe.[283] Quetelet developed the statistical concept of the "average man" and sought to discover, through statistics and the study of "social physics," the explanation of societal behavior. Needless to say, his observations and conclusions were controversial.

In 1829 Quetelet published his first statistical study concerned with commerce, education, crime, and other characteristics of population, and, though he made abundant reference to Dupin's *Forces commerciale et productive de la France*, he made no attempt to make a map, even though he was much concerned with regional variations in the Low Countries.[284] In 1831 he presented a *mémoire* on the *penchant au crime aux differens âges* to the Académie Royale des Sciences et Belle Lettres at Brussels which was published separately that year and in 1832 by the Académie.[285] In it he included three *cartes figuratives* of the region of France, the Low Countries, and the Duchy of the Lower Rhine: plate 1 of instruction, and plate 2 of crimes against property and crimes against people (FIG. 77). He compared his statistics of instruction with those of Dupin and believed his treatment to be better.[286] He also observed that the printing of the *mémoire* was almost completed when he received a letter from Guerry "whose name is well known for various works in statistics, and especially . . . on the statistics of crime which he has published with M. Ad. Balbi."[287]

Quetelet's method of portrayal is fundamentally different from that of Dupin and Guerry. Those first two thematic maps of moral statistics are choropleth maps, which were themselves innovative. Quetelet's treatment, equally innovative, was an attempt to show a smoothed, continuous distribution with tones proportioned according to the principle that the darker the greater (ratio of crimes), such as would result if one were to use today's familiar dot-map technique in which each dot would represent a given number (of crimes).[288]

The cartographic portrayal of moral statistics was now in full swing. In 1833 Guerry published an essay on the moral statistics of France resulting from a *mémoire* presented to the Académie in 1832 showing maps and giving tables of data concerning:
    1. Crimes against people
    2. Crimes against property

Maps of the Social Environment

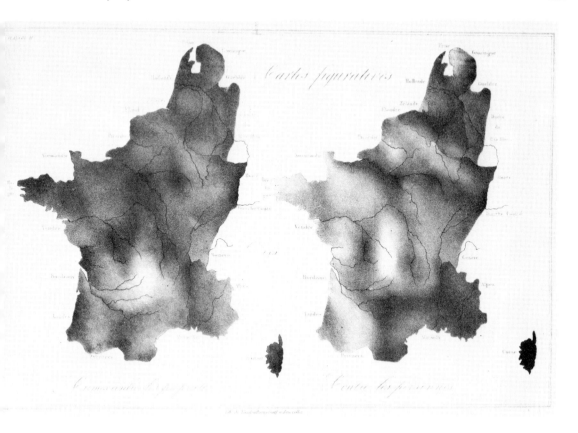

**Figure 77** Quetelet's 1831 plate 2 "Cartes figuratives—Crimes contre les propriétés—Contre les personnes." Original 335 × 230 mm. Lithograph, crayon shading. (Courtesy of the Center for Research Libraries, Chicago.)

    3. Instruction
    4. Illegitimate children
    5. Donations to the poor
    6. Suicides.

He cited Humboldt and Playfair as justification for the graphic portrayal and referred to Quetelet's ...*penchant au crime*... of 1831.[289] In 1837–38 Guerry joined Dupin as a foreign honorary member of the Statistical Society of London. Guerry's study was widely read and, according to his biographer, created a sensation. It was the subject of numerous comments and at least three papers given at meetings of the Statistical Section of the British Association.

One of the papers was by W. R. Greg, in 1835, who wanted to know how statistics of other countries compared with Guerry's results. He found that statistics for "our own country" were inadequate and selected the Netherlands. He considered the Netherlands under the same headings as Guerry "and he *mapped* these several heads as M. Guerry has done."[290] Greg thought he had discovered a relationship, namely that the proportion of births to a marriage appears to vary inversely as the proportion of marriages to the population.

In 1835 Quetelet published his most influential study on man and the development of his faculties, which made a very great impression over the whole of Europe through criticism, republications, and translations.[291] Although the references in the text to the three maps are word for word as in his 1831 *mémoire*, the maps were remade, but they are essentially the same. The book was printed, without authorization, in Brussels in 1836, then in 1838 it was translated with Quetelet's blessing into German and the maps were redrawn.[292] In 1842 an English edition appeared, also with Quetelet's permission, and the maps were redrawn again.[293] Each time the maps employed the same proportional shading technique. Although there was, of course, no photographic process involved, the reproductions are quite faithful. Because of the wide diffusion of Quetelet's maps in the French, German, and English editions of his book, his innovative smooth, proportional, shading technique was widely employed during the next quarter-century for portraying relative densities.

Whereas Guerry and Quetelet had been more interested in the interrelationships among the distributions of education and crime, others were equally interested in portraying the regional differences. One such was d'Angeville, who published a series of sixteen thematic maps in 1836 to accompany his lengthy essay on the population of France, in which he pointed out that the "shaded maps are a graphic means to make up for the aridity of the numbers."[294] His maps covered a wide range of subjects, such as longevity and physical constitution, as well as the usual moral statistics of instruction, crime, and so on. D'Angeville ranked the *départements* in terms of the subject—for example, the number of people per each person accused of a crime during the five-year period 1828–32—and grouped the ranks in five classes. He put the rank numbers of the eighty-five *départements* on the map and then shaded the map with five tones, each tone covering one class of seventeen *départements*. Each map also carries a table showing the rank order and the actual statistic for each *département*, plus Corsica (FIG. 78).

The census of Ireland taken in 1841 was accompanied by several maps. One, a population map, was noted in the last chapter, and others will be discussed and illustrated later in this chapter. We take note here that the census included:[295]

> For the purposes of conveying to the eye a general representation of the degree in which total igno-

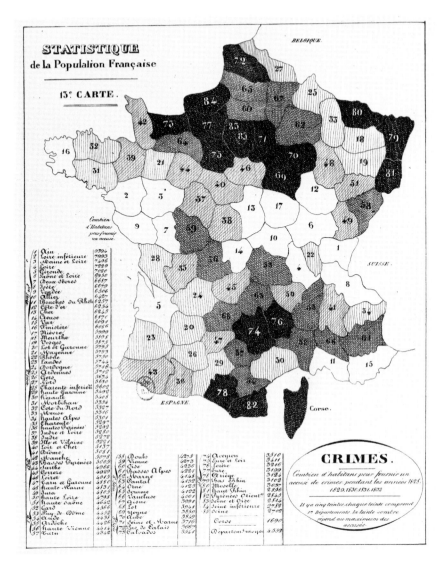

**Figure 78** D'Angeville's 1836 map of the number of persons per each one accused of crime in the years 1828–32. The greater the ratio of accused, the darker. Original 187 × 239 mm. Lithograph. (Photo. Bibliothèque Nationale, Paris.)

rance of elementary instruction still prevails in the several counties we annex a shaded Map (Plate 3), by which it will be seen that, in a large portion of the island, there is still considerable darkness.

One hopes that the cartographic pun was intentional. In design and execution the map is the same as FIGURE 54.

The next major cartographic work on moral and educational statistics was by Joseph Fletcher, who from an early age had been engaged on reports and works having to do with the health and occupations of people. Born in 1813, he was educated as a barrister, served as a secretary to the "handloom enquiry commission" and the "children's employment commission," and was an inspector of schools. In 1847 he published a paper, accompanied by only one simple reference map, but including many tables regarding educational and moral statistics. He asserted that the map served as key and that:[296]

> A glance down the vertical columns of these tables will convey all that could be pictured forth by an expensive series of shaded maps, showing the relative intensity of each element.

His anticartographic attitude soon did an about-face. Only two years later, in 1849, he published another study, but this time it was accompanied by twelve relatively elaborate maps.[297] He apparently had reacted promptly to suggestions from the highest quarter for he wrote:[298]

> A set of shaded maps accompanies these tables, to illustrate the most important branches of the investigation, and I have endeavoured to supply the deficiency which H.R.H. Prince Albert was pleased to point out, of the want of more illustrations of this kind.

Fletcher's maps are reminiscent of d'Angeville's in that for each subject the counties are ranked and the rank number is placed on the map (FIG. 79). But on all but one map the range is divided into seven classes with seven tones, according to the proportions above and below the average of England and Wales, so that, unlike d'Angeville's maps, a category of shading might not occur. Each map is accompanied by two tables listing counties with their statistics, one table in alphabetical order, the other in rank order. The following data for England and Wales are mapped:

1. Dispersion of the population: 1841
2. Real property in 1842 in proportion to the population: 1841
3. Persons of independent means in proportion to the population: 1841
4. Ignorance, as indicated by the men's signatures by marks in the marriage registers: 1844
5. Crime, as indicated by the gross criminal commitments of males to Assizes and Quarter Sessions: 1842–47
6. Commitments, for the more serious offenses against the person and malicious offenses against property: 1842–47

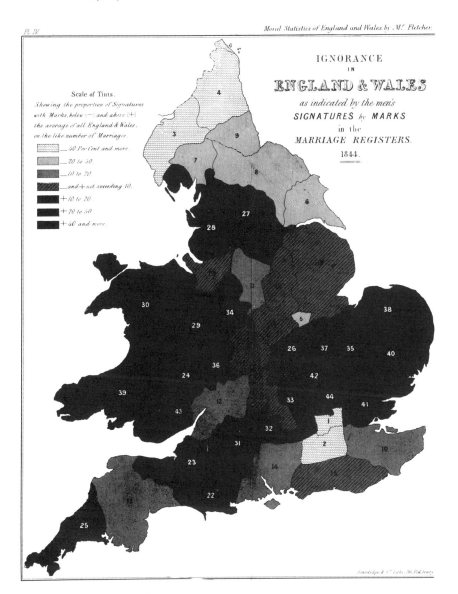

**Figure 79** Fletcher's 1849 map of "Ignorance... as indicated by the men's signatures by marks in the marriage registers: 1844." Original 190 × 235 mm. Lithograph. (Courtesy of the University of Wisconsin Memorial Library.)

    7. Commitments, for offenses against property, excepting only the "malicious": 1845–47
    8. Commitments, for assaults and miscellaneous offenses of all kinds: 1842–47

9. Improvident marriages (those of males under twenty-one being so designated): 1844 and 1845
10. Bastardy, as indicated by the Registry of Births: 1842 and 1845
11. Pauperism, as indicated by the proportion to the whole population of the persons in the quarter ending Lady-Day: 1844
12. Deposits in the savings banks in proportion to the population: 1844

Fletcher devotes considerable space to comparisons of one map with another searching for hypotheses to "explain" the variations in the amount of crime and such. He pays homage to both Quetelet and Guerry, but he disagrees with the latter's assertion that there is no general coincidence between the moral features of different districts and the varying amounts of technical instruction prevailing in them.

Numerous studies of moral statistics appeared after midcentury and the techniques for displaying such data on thematic maps were rather well established by then. It is appropriate to end this section by taking notice of two significant studies that used maps extensively. The earlier appeared in 1862 as volume 4 of Henry Mayhew's extraordinary *London Labour and the London Poor*. Mayhew, a journalist-sociologist and a founder of the magazine *Punch*, wrote a voluminous account of the relatively unsavory social conditions of the time, and in an appendix to the last volume he gives fifteen "maps and tables illustrating the criminal statistics of each of the counties of England and Wales in 1851."[299] The variety of subjects is very large, ranging from the number of early marriages to committals for bigamy to attempts to procure miscarriage.[300] One map and set of statistics concerns the number of females to every one hundred males, and one wonders how that can be construed as a "criminal statistic." The maps are simple, consisting of county outlines with the names and averages (FIG. 80). Those counties above the average are left white, while those below are shown black.

Guerry, who pioneered in the mapping of criminal statistics, produced a set of fifteen sophisticated maps and graphs in 1864, two years before his death.[301] With paired maps, he compared the moral statistics of England and Wales with those of France on the subjects of crimes against the person, against property, murder, rape, larceny by servants, arson, instruction, and (for France only) suicide. The maps are relatively sophisticated in appearance, neatly drawn and shaded, and each is accompanied by a table with the counties and *départements* in rank order, the ranks being noted on the maps (FIGS. 81 and 82). They "created a sensation in Germany, England and the United States," and in 1864 Guerry was sponsored at the meeting of the British Association for the Advancement of Science at Bath by William Farr, the statistician,

*Maps of the Social Environment*

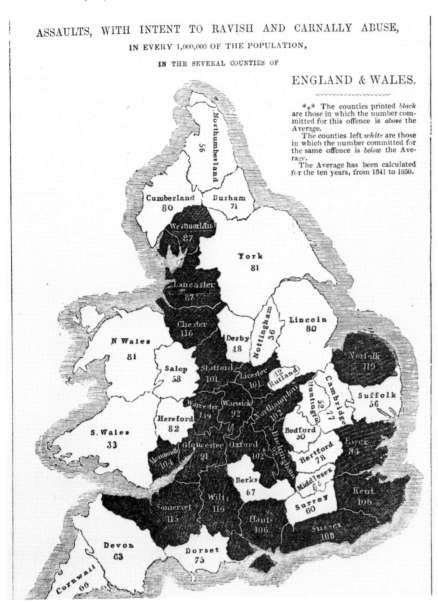

**Figure 80** Part of the map showing the number of persons per 1,000,000 committed for assaults, Map 12 in Mayhew's *London Labour and the London Poor* . . . . Reduced 10 percent. Wood engraving. (Courtesy of the British Library.)

who at that time was Treasurer of the Statistical Society.[302] The maps of Guerry's atlas were displayed in the meeting room, and he was delighted to be asked to comment on them.

**Figure 81** Guerry's 1864 map of crimes against property, plate 3 in his *Atlas*... accompanying *Statistique morale de l'Angleterre comparée avec... la France*. Original 240 × 355 mm². Lithograph, printed two colors. (Courtesy of the British Library.)

Maps of the Social Environment

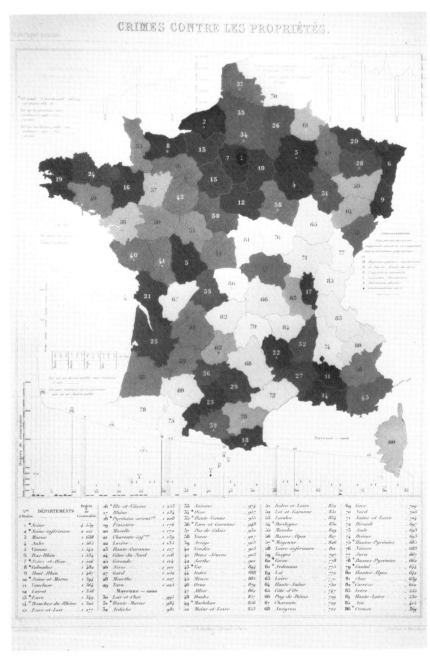

**Figure 82** Guerry's 1864 plate 4 matching map of France. See FIGURE 81.

Although thematic maps of moral statistics continued to be made, especially of instruction, their developmental period had run its course. Subsequent attention seems to have been oriented more toward sociological interpretation and away from geographical variation. In a sense this reflects a greater concern with Quetelet's provocative ideas of "social physics" and a lessening of interest in the investigations of regional differences, such as by Dupin, Guerry, and Fletcher.[303]

## Medical Maps

Some of the curiosity aroused during the period when thematic mapping began was stimulated by the epidemics of yellow fever and, especially, cholera which latter swept into western Europe repeatedly in the early and mid-nineteenth century.[304] Medical mapping invites a broad definition, since probably even before the time of Hippocrates people associated environmental conditions with sickness. Thus, even maps of swamps can be classed as "medical maps." Similarly, maps of hospitals, routes of spread, quarantine cordons, and even the placing of ambulances on a battlefield can come under the rubric.[305] If we restrict our attention to those maps which can reasonably be called thematic, we narrow the field considerably, since then the interest focuses on maps which attempt to show the structure of the distribution of a disease. In some cases this will include maps of routes of spread, made to help people understand the mechanisms involved, but mostly they show geographical incidence or frequency and were prepared in order to contribute to an understanding of causes. In the early period of thematic mapping of disease, yellow fever came first, and the first maps of that disease were "spot maps" showing locations of occurrence, the earliest apparently being 1798.[306] However, cholera was the major subject of interest in western Europe, since this devastating, dehydrating illness was epidemic and therefore unusual, compared with the many endemic diseases, such as tuberculosis, which were equally devastating over long periods, but which excited no particular interest. The natural fear of the unusual was enhanced by the fact that by this time it was assumed that there was some physical cause for cholera, rather than its being a visitation, and that somehow geographical variation had something to do with it.[307]

A variety of cholera maps began to appear soon after the disease became pandemic in India in 1817, and by 1832, when it had spread over Eurasia and North America, a large number of maps were made. Through the year 1832, Jarcho lists twenty-nine cholera maps published in the twelve years after 1820, ten in 1831 and fourteen in 1832.[308] Most of these maps showed routes of spread, dates, and spots or regions of occurrence. They became somewhat more sophisticated cartographically when variable shading began to be used to show relative

*Maps of the Social Environment*

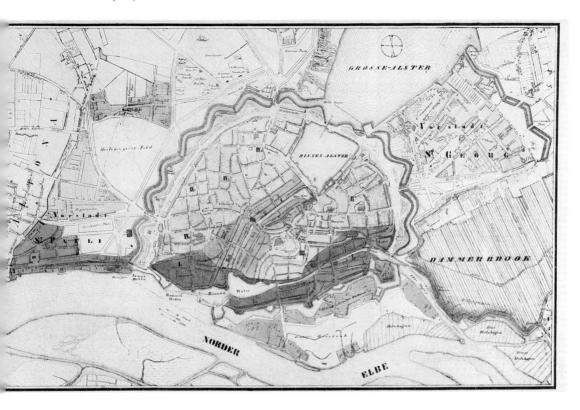

**Figure 83** Rothenburg's 1836 map accompanying his *Die Cholera Epidemie des Jahres 1832 in Hamburg*. Original 446 × 284 mm. Lithograph, hand colored. (Courtesy of the British Library.)

intensity or density. One of the earlier of such maps was Rothenburg's map of the cholera epidemic in Hamburg in 1832 (FIG. 83).[309] In his analysis of the epidemic Rothenburg points out that his account of the places of occurrence and mortality are tedious and that his map might be more interesting "where the gradations of the greater or lesser strength of the epidemic are given by the stronger or weaker shadings of the red color."[310]

Rothenburg's map was redrawn and included in the report of the British General Board of Health on the epidemic of cholera which for the second time visited Britain in 1848–49. It was included because it "places in a striking point of view the predominating influence of locality over the progress of the disease."[311] Three similar maps were included in the report, but the shadings were not keyed to a legend.

It is probably a safe generalization that many of the large number of maps made to illustrate the distribution of cholera cases were manuscript and were displayed when reports were given but then never reproduced. Book librarians, for very good reason, class maps as "fugitive materials," and, because they are bothersome to store, in the course of time they disappear. No doubt many such manuscript maps were prepared with care, and because they were meant to be displayed they were likely quite communicative, especially because color can be easily used on manuscript maps.

A good example of the foregoing is provided by the map prepared by Robert Baker, district surgeon in Leeds, in his report to the Leeds Board of Health concerning the cholera epidemic during May to November 1832, when more than seven hundred people died from the disease. In his report he states:[312]

> Accompanying this Report, are a Map and Schedule, on the former of which are depicted the streets, sewered and paved by the town—(colored blue,)—by private individuals—(yellow,)—the townships (green). The strong black lines denote the river and water-courses; the smaller lines in the middle of the streets point out the common sewerage, and, of course, by the colour of the street, it will be perceived whether it is public or private. The red colour marks only the districts in which the Cholera prevailed.

It must have been a colorful map with blue, green, yellow, and red. The printed map with the published report is not that map, but a small (275 × 216 mm), crude, black-and-white lithograph on which the reference states the "Cholera parts Lined & colored RED." Presumably someone was to hand color the lined areas.[313] The printed map is clearly not the map displayed by Baker.[314]

Another example of a lost map is that referred to in a paper read to the Statistical Society of London in 1835 regarding the incidence of cholera in 1831–33.[315] In it the author, in describing the sources of his facts, observed that:

> A map was constructed by Sir William Pym, Superintendent General of Quarantine, showing at one view all the places in Great Britain which had been attacked by the disease. Sir William also appended to the map a voluminous Index or Table.... These curious and important documents were lodged in the Royal Library by His Majesty's most gracious desire.

Even royal interest did not suffice: the report exists, but the map is missing. It would not be surprising if many such interesting thematic maps disappeared in western Europe in similar fashion. The concern and effort to understand the cholera outbreaks must have caused many

Maps of the Social Environment

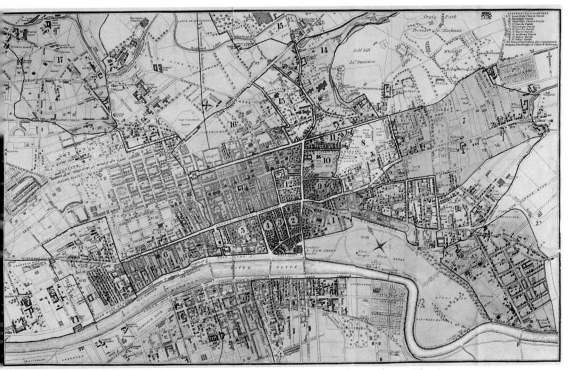

**Figure 84** Perry's 1844 map of Glasgow showing the relative prevalence of an epidemic; the darker the tint the greater the incidence. Original 410 × 270 mm. Lithograph, hand colored. (Courtesy of the British Library.)

such maps to be prepared for the various boards of health and governing bodies of cities, in the hope of ascertaining the environmental factor or factors affecting the occurrence of cholera.[316]

Fortunately, some of the maps have been preserved, and the accompanying descriptions of the appalling conditions of filth and crowding in which lived the masses of people who had moved to the industrial cities, told in the direct words of frustrated physicians, are eloquent enough to make one nauseated. Sometimes the reports and surveys called for considerable organization and planning.

One such is the report by Dr. Robert Perry, senior physician to the Glasgow Royal Infirmary.[317] Perry asked each of the district surgeons in Glasgow to provide, in addition to the weekly report of new cases, a map of his district showing the regions of major outbreaks of an epidemic (probably of influenza) which began in December 1842. When he had all the data, he compiled it on a map of Glasgow (FIG. 84). It was not always easy, for one of his surgeons wrote:

> I have not yet procured a map convenient enough to jot the localities where the disease mostly prevailed, but will have it prepared in a short time.

Dr. Perry then:

> laid down and numbered the different districts on a map of the city... marking with a darker shade those parts where the epidemic was most particularly prevalent, which shows that those places most densely inhabited, by the poorest of people, have suffered most severely.[318]

Dr. Perry's thematic map was innovative in one respect, to say the least, since the printing of the report, the coloring of the map, and so on, were:

> wholly the work of the inmates [of the Lunatic Asylum at Gartnavel]. It is hoped that this circumstance will induce the reader to make allowance for anything he may observe amiss in the manner in which this, their first essay, has been performed.[319]

The interest in epidemic cholera was so great that it is probable that cholera maps far outnumbered other medical maps, yet others were made in considerable variety. We can only note a few. The range of subject matter is great. For example, C. F. Weiland, an active thematic cartographer for his period, in 1835 made a map of various mineral waters.[320] Of much more interest is a map made in 1839 of the distribution of hernia in France (FIG. 85).[321] Its author, J.-F. Malgaigne, a surgeon, obtained statistics of the incidence of hernia as found during the examination of military recruits in 1836 and 1837. He used the map to test tentative hypotheses that incidence of hernia tended to be high in valleys and also high in regions where cider was dominant over wine as the standard drink, as well as a suggestion by others that ingestion of olive oil favored hernia. He drew these critical lines on his map and found he had to discard all the hypotheses.

The investigation of possible correlations among environmental factors and human abnormalities is limited only by the imagination. One such nosogeographic study in 1843 was made by E. H. Michaelis, who investigated the incidence of cretinism and the proportion of deaf-mutes in the population as related to elevation and other environmental characteristics in Canton Aargau.[322] The map is curious for reasons beyond the subject matter, since Michaelis employed shaded (cf. illuminated), contours to show elevations, and devised a complex symbolism to try to portray the medical data (FIG. 86). The diameters of the circles in the red rectangles represent the relative frequency of cretins, while the lengths of bases of the rectangles show the relative frequency of deaf-mutes, both based upon the ratios of these occurrences with the populations of the adjacent villages. These numerical

*Maps of the Social Environment*

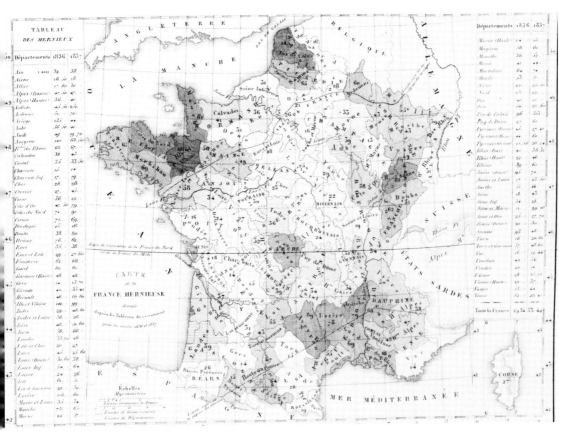

**Figure 85** Malgaigne's 1840 "Carte de la France Hernieuse." Original 228 × 167 mm. Lithograph. (Courtesy of the University of Wisconsin Health Sciences Library.)

values are also shown as red digits, those for cretins being placed above or below the rectangles and those for deaf-mutes alongside it.

In a paper read before the Statistical Society of London in 1853, Dr. J. R. Hubertz of Copenhagen exhibited a map of Denmark colored geognostically, showing by digits the number of deranged persons per one thousand total population.[323] Examination of the early issues of the numerous journals published by the many statistical and medical societies that came into being in the nineteenth century in western Europe would no doubt be rewarding.

Probably the most famous medical map of this or any other era is the dot map prepared by Dr. John Snow to accompany the second edition of his prize-winning essay on the communication of cholera (FIG. 87).[324] The placing of dots on a map to mark the locations of individual cholera deaths was not new,[325] but Snow's inductive use of the technique was.

**Figure 86** Part of Michaelis's 1843 map of the distribution of cretinism in Canton Aargau. Original scale 1:125,000, reduced 50 percent. Lithograph, printed color. (Courtesy of the British Library.)

Whereas such investigators as Malgaigne, Michaelis, and Hubertz had mapped general distributions to observe possible coincidences among general distributions, Snow employed the thematic technique to show "the topography of the outbreak" and how proximity to a single source of pollution affected the frequency of mortality. Writing of the Broad Street, Golden Square, area in London in August–September 1854, Snow related:

> Within two hundred and fifty yards of the spot where Cambridge Street joins Broad Street, there were upwards of five hundred fatal attacks of cholera in ten days.

There would have been more except for the flight of the population. Although Snow did not know what actually caused cholera, the structure of the distribution of deaths as revealed by the mapping convinced

Maps of the Social Environment 177

**Figure 87** Snow's 1855 map 1, showing the location of cholera deaths in relation to the Broad Street pump (near center) and to the other public pumps in the vicinity. Original 415 × 384 mm. Lithograph. (Courtesy of the British Library.)

him that water from a particular well was the local source. On the seventh of September Snow persuaded the authorities to remove the handle of the public pump at the corner of Broad and Cambridge streets. Deaths from and new cases of cholera declined immediately. His memory is retained locally in the name of a public house, "The John Snow," at the site of the Broad Street pump, which is more than a little incongruous, since he became a total abstainer at age seventeen. (FIG. 88.)

**Figure 88** The John Snow public house at the corner of present Broadwick and Lexington Streets, the site of the Broad Street pump. (Author's photograph.)

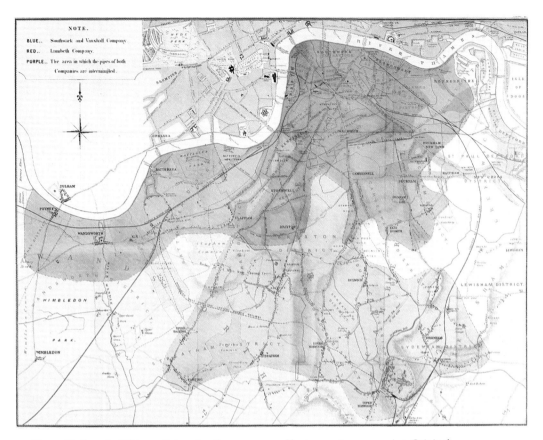

**Figure 89** Snow's 1855 map 2 showing the areas served by two water companies. Original 535 × 410 mm. Lithograph, printed color. (Courtesy of the British Library.)

Snow used thematic cartography in another way to demonstrate his contention that polluted water was the carrier of cholera (FIG. 89). Before 1852 two water companies, the Southwark and Vauxhall Company and the Lambeth Company, supplied the area of London south of the Thames, both obtaining their water from that heavily polluted river. In 1852 the Lambeth Company changed to a source free of London's raw sewage. Dr. Snow's analysis of the number of deaths from cholera in the areas supplied by the two different companies showed that the area receiving polluted water had a ratio of 71 deaths per 1,000 houses, but the number in the area receiving unpolluted water had dropped to only 5.[326]

Several other medical maps should be mentioned, of the many made during the formative years after 1830. Engelmann reports that F. Schnurrer in 1827 had prepared a colored map showing the geographical spread of sickness for presentation to a meeting of natural scientists and doctors in Munich.[327] There had been earlier world maps of cholera, but it remained for Berghaus to produce in 1848 the first map in an atlas showing the geographical distribution of human diseases (FIG. 90).[328] Johnston did not use that map in the first edition of *The Physical Atlas*, but in the second edition (1856) the last map, number 35, shows the geographical distribution of health and disease in connection chiefly with natural phenomena.

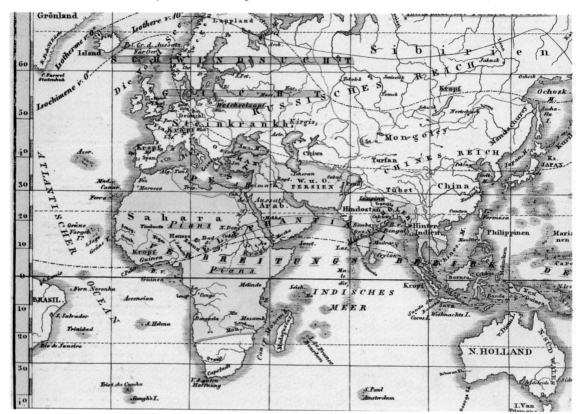

**Figure 90** Part of the 1848 world map showing the geographical spread of the principal illnesses to which man is exposed, seventh Abteilung, Anthropography, no. 2, in Berghaus's *Physikalischer Atlas*. Reduced 10 percent. Copperplate engraving, hand colored. (Courtesy of the British Library.)

Quite different from the other medical maps is the cholera map of the British Isles prepared by Petermann soon after his move to London in 1847. The map sheet, undated but known to be 1848, showed England, Scotland, Wales, and Ireland and used the shading technique employed earlier by Quetelet (see FIG. 77) to show the variation in mortality—the darker the greater (FIG. 91).[329] Places "attacked by the disease" are

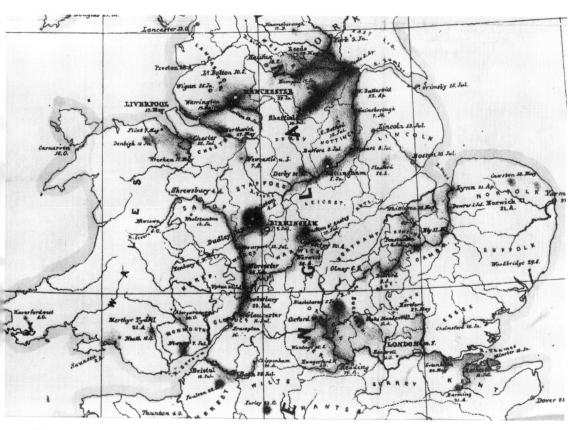

**Figure 91** Part of Petermann's 1848 "Cholera Map of the British Isles showing the Districts attacked in 1831, 1832, and 1833." Full size. Lithograph, crayon shading. (Courtesy of the British Library.)

named, as well as the dates when the first case occurred at that place. Petermann prepared some statistical notes to accompany the map in which he discussed, among other things, the value of geographical delineation.[330] He also included a small map of London, together with a table listing registration districts, population, cases, deaths, and the ratio of deaths to population. The map is tinted red, the darker shades being applied where the mortality ratio was higher.[331]

Petermann's cholera map may well be the first map he prepared when he moved to London from Edinburgh to found his own business, and it may also be among his earlier attempts at employing lithography rather than the copper engraving he had been primarily doing while with Berghaus and Johnston. That when he struck out on his own he turned to a medical map indicates the degree of interest in that subject, so appropriate for thematic mapping.

### LIVING CONDITIONS

Many other aspects of the social environment became subjects for the early thematic cartographers, and one of the earliest of these was living conditions. This was partly because of general concern over the quality of accommodations and partly because housing characteristics were subject to taxation. For example, in England the tax on fireplaces and chimneys was replaced in 1766 by one on the openings of houses that were apparent from the outside. A similar tax in France was imposed at the time of the Revolution.[332] In 1836 d'Angeville, in his sixteen maps concerning the French population, included one map of the number of doors and windows per 100 inhabitants in 1831 (FIG. 92). As in all his other maps, the ranked *départements* are grouped in five classes, the poorest category being made the darkest. The best-endowed *département*, Eure, had 197 openings per 100 inhabitants, while the poorest, Côtes du Nord, had only 61.

The census of Ireland taken in 1841 adopted a similar but more elaborate approach to rating accommodations.[333] Each house was ranked by the number of rooms, the number of windows, and the quality of the construction materials, and several classes in each category were established. Each house was given a class number from each category; these were then summed, resulting in four classes of houses. Accommodations were then ranked according to the class of the house and the number of families occupying it: first class was a first-class house with one family; fourth class ranged from a fourth-class house with one family to a first-class house with more than five families. The commissioners

> with a view toward giving a general graphic representation of the relative prevalence of the worst class of lodging, we annex a map (Plate 2), shaded in the usual way, for that purpose.

On the map the percentage of families occupying fourth-class accommodation is entered in each district and the higher the number (i.e., the poorer) the darker the shading (FIG. 93).

The 1841 census of Ireland was generally very well received in all quarters. The *Nation* for 16 August 1845 carried a very full and complimentary review, devoting almost one whole column to the maps. It commented that "To the mass of people the maps of Ireland would be the most familiar teachers." Larcom, generally conceded to be the architect of the census, reported on it to the Statistical Section of the British Association for the Advancement of Science meeting at Cork in 1843.[334] The reporter for the Statistical Society of London went so far as to write:[335]

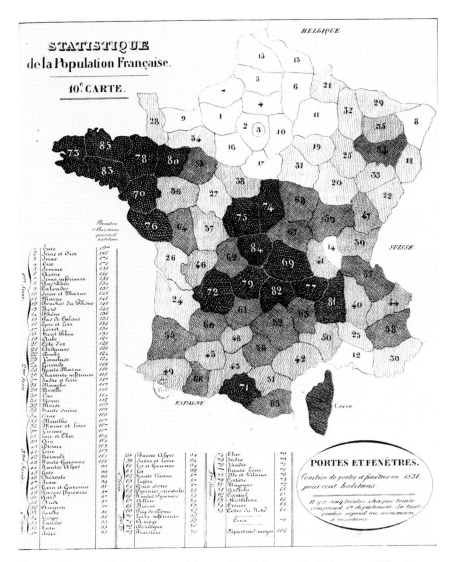

**Figure 92** D'Angeville's 1836 map of the number of doors and windows per 100 residents in 1831. Original 187 × 239 mm. Lithograph. (Photo. Bibliothèque Nationale, Paris.)

The most important report... was that of the Census of Ireland in 1841... it may serve as a model worthy of consideration at a future census, even in this country. It was accomplished by illustrative shaded maps of the population, education and crime.

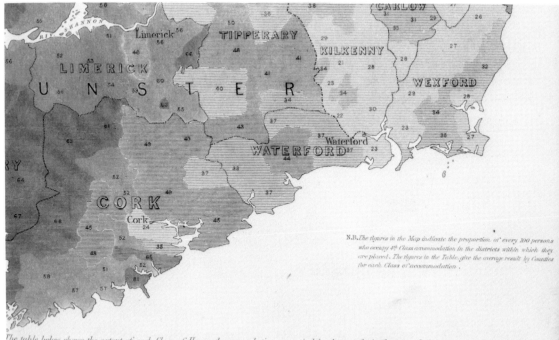

**Figure 93** Part of the map showing the percentage of persons occupying the poorest accommodation, plate 2, *Census of Ireland, 1841*. About full size. Copperplate engraving. (Courtesy of the British Library.)

The pattern of the census of Ireland was indeed followed to some extent by other censuses, but not entirely. One map did not receive universal acclaim (FIG. 94). In a special report on Dublin, W. R. Wilde included a "sanitary map" of the city in which he bravely classed and colored the streets as follows:[336]

    1st Class, private streets
        high class residence                                              Blue
    2d Class, private streets
        more unhealthy nature, some shops, etc.          Crimson

Maps of the Social Environment

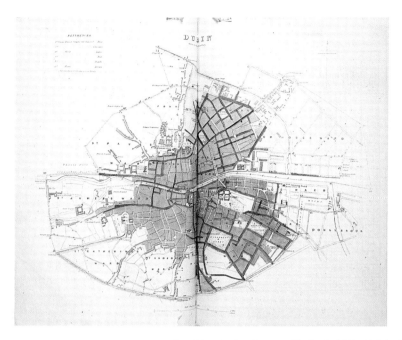

**Figure 94** Wilde's sanitary map of Dublin in the *Census of Ireland, 1841*. Original scale 1:253,400 (4 inches to 1 mile). Copperplate engraving, hand colored. (Courtesy of the Library of Congress.)

    1st Class, shop streets                                                        Yellow
    2d Class, shop streets                                                     Red
    3d Class, shop streets
        inhabited by lowest class of traders,
        artizans, huxters, low population                     Purple
    3d Class, mixed streets
        small shops, including middle class
        private residences, etc.                                  Brown

The map was reviewed in the *Athenaeum* of London as follows:[337]

> The prevailing notion has been, that there has been a little too much in Ireland of distinctions by colour. The census-makers, however, have thought differently on this subject, for we observe that they have actually classified the streets of Dublin by different tints of blue, crimson, yellow, red, purple, and brown. This will produce infinite discord and endless confusion. Those who live in a blue (or first-class) street, will turn up their noses at those who dwell in a crimson (or second-class) street; the inhabitants of crimson streets will look down on the inhabitants of yellow

streets..., we may confidently anticipate a rupture of all social relations. The influence of colours is prodigious, and they are nowhere so influential as in Ireland. We have no doubt that this egregious blunder in map-making will set ten thousand families, in Dublin, at loggerheads. The blue streets may remain on terms of cold acquaintanceship with the crimson, but it will soon be absolutely necessary for the crimson to give the yellow the cut direct. It will then be blue and crimson against the other hues.

We bring this chapter on the thematic mapping of the social environment to a close with two maps. The first is a map prepared to accompany and illuminate a report to the Poor Law Commissioners in Britain. The costs of poor relief had been steadily rising, and a centralization of authority had become law in 1834. The Poor Law Board, among other things, was concerned about the regulation of sanitary conditions, and an act of 1837 required for the first time the official registration of births, marriages, and deaths, which made available the statistics used by such investigators as Fletcher and Mayhew, whose maps we have already noted. Edwin Chadwick, the secretary of the board, published in 1842 his report on the sanitary conditions of the labouring classes.[338] This contained two maps, one of Bethnal Green and the other of Leeds (FIG. 95). Both maps showed classes of housing, and data on the distribution of the incidence of disease. The Leeds map "which Mr. Baker has prepared at my request" (the same person as Baker 1833) shows the locations of the various kinds of textile mills and distinguishes the "less cleansed districts." One may assume that thematic maps of a similar genre appeared in official reports and studies in other countries as the moves to better the shocking living conditions of most of the people in western Europe gathered impetus throughout the remainder of the nineteenth century.

Although the last map lies beyond the general time frame within which thematic cartography grew to maturity, it is included to illustrate the development that took place in these kinds of maps in the approximate half-century since they were first made. In 1889 Charles Booth, shipowner and sociologist, published the first volume of the seventeen-volume work *Life and Labour of the People in London*.[339] In it were several maps, but the poverty map of East London is almost a direct descendent of Wilde's sanitary map of Dublin, since the streets are also classified, though a bit more strongly (FIG. 96). The lowest class, black on the map, is designated as "very poor, lowest class...vicious, semi-criminal."

Maps of the social environment continued to have wide distribution and became a regular part of atlases and population studies.[340] They

Maps of the Social Environment

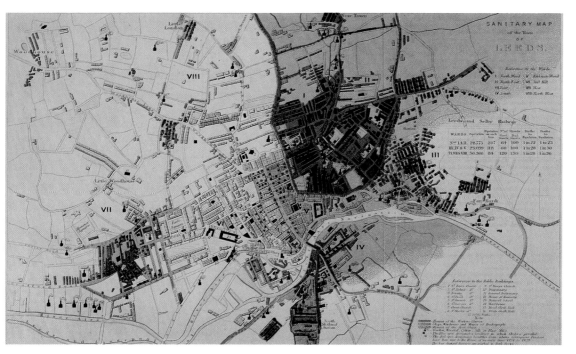

**Figure 95** "Sanitary Map of the Town of Leeds" in Chadwick's 1842 *Report on the Sanitary Condition of the Labouring Population* .... The darker the brown, the "less cleansed" the district. Original 290 × 172 mm. Lithograph, printed color. (Courtesy of the British Library.)

reflect the wave of reform that swept over western Europe during the nineteenth century. The mix that brought them into being ranged from strictly practical problems, such as epidemics and the costs of poor relief, to the interest in the theory that man's characteristics and behavior, en masse, ought to be as predictable and as subject to laws as were the inanimate aspects of the environment. Curiosity knew no bounds. For example, in England an 1849 report by a Select Committee on Public Libraries included a map of Europe showing the number of books in libraries in each of twelve national areas per 100 of the population.[341] The best endowed were the "smaller German states," with nearly 450 books to every 100 persons; the poorest were Holland and the British Isles, with from 53 to 63. The portrayal of all these kinds of information required a new approach to dealing with data and their symbolization, and, as we shall see in the next chapter, fortunately the technical developments of the period allowed it. The cartography which had been essentially the same since the introduction of printing at the end of the fifteenth century now had a new area of subject matter added

188  Maps of the Social Environment

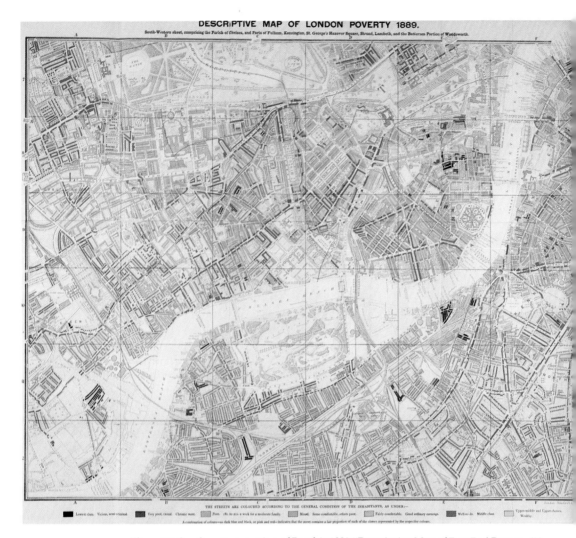

**Figure 96** Southwestern section of Booth's 1889 "Descriptive Map of East End Poverty," in *Life and Labour*, Vol. 1, East London. Reduced 75 percent. Lithograph, printed color overprint on gray base. (Courtesy of the British Library.)

to its repertoire. The only certainty on earth is change, sometimes slow and sometimes fast. By historical standards, the development of thematic cartography was rapid.

# Seven

# NEW TECHNIQUES AND SYMBOLISM

THE EFFECTIVE MAPPING OF MANY OF THE SUBJECTS PORtrayed on the early thematic maps required new symbolic methods. For centuries most of the prethematic, general maps had relied on point symbols to show the locations of places, on line symbols to show coasts, roads, rivers and boundaries, and on stippling, hatching, shading, or color area symbols to portray water bodies. Varied representations were devised to show where hills and mountains existed, ranging from flat colors to side views of "sugar loaves" or "mole hills," to exceedingly complex delineations, often along with an equally varied symbolization of forested areas.[342] Forests and mountains were about the extent of "thematic" portrayal on maps before the time when interest turned to variations within classes of data.

As I noted before, the new concerns and outlook began to emerge in the latter part of the seventeenth century, and even in that early period a fresh approach to graphic portrayal was called for. Happel and Kircher needed to show the movements of water (FIGS. 11 and 29), Halley required some way to portray wind movements (FIG. 21), and some system was necessary to delineate differences in compass declination. One must keep in mind that most of the data that were mapped up to the latter part of the eighteenth century were conceived as being basically qualitative—that is, consisting of variations in kind or character as opposed to being quantitative. Even compass declination is a rather qualitative concept which, fortunately, could be designated by a numerical system as well as with bearing designations. The era of counting, from population censuses to the accumulation of economic data, which characterizes the latter part of the period of thematic development, raised a host of representational problems that earlier had not even occurred to cartographers. How could one portray the notion of numerousness, of more in this region than in that? The concept of density—that is, so many of something per unit area—needed to be shown, as well as other kinds of ratios and rankings. The movement of various amounts of items from one place to another also required new systems of delineation.

In a broad sense the development of cartographic technique fits the maxim "where there's a will there's a way," since, almost as soon as problems arose, solutions were tried out. As we shall see, some of the earlier attempts, though imaginative to say the least, were not very successful graphically. On the whole, however, the trial-and-error approach proved relatively efficient. This was partly because the thematic cartographers of the nineteenth century were an ingenious lot, but also it stemmed from the fact (with some notable exceptions) that there was excellent communication among them. Most were familiar with what others were doing. Letter writing apparently was common, and the host of journals that sprang up were widely circulated within the intellectual community of western Europe.

The significant, formative years in the development of portrayal techniques in thematic cartography were the first five or six decades of the nineteenth century. During that period most of the methods in use today were invented, tried out, modified, and finally settled upon. This is not to say that nothing innovative has occurred since then. Considerable sophistication in conceptual approach and in processing data has definitely been achieved. Some new methods of portrayal have been instituted, some successful, such as linkage displays where complex sets of movements are shown by connections of origins and destinations, and some not so successful, such as the attractive but misleading proportional spheres. Nevertheless, generally speaking, the great majority of the techniques used in thematic cartography today were devised and put to use before 1860. The list is long: proportional point symbols, the line of equal value (isoline), the choropleth and dasymetric methods, sectored circles or pie graphs, variable shading, the dot method, class intervals, and flow lines. Some of the first uses of the methods were buried in out-of-the-way places and had to be reinvented, some developed with controversy, and some came into being fully developed on the first try, so to speak. In this chapter I will examine the origins of some of the representational innovations in thematic cartography, but before turning to them, it is appropriate to review an intimately related facet of the cartographer's task: how to duplicate the desired delineation.

### Techniques of Duplication

Throughout the long history of cartography the manuscript mapmaker has had a relatively easy construction task, being limited in his graphic delineation only by his equipment and his repertoire of skills. He may use any colors, line weights, inks, shadings, and so on, that he wishes, and if he is skillful he can therefore make a very effective map. On the other hand, he pays for this freedom in that if he wants a copy it must all be done over again, laboriously, by hand. Even before movable type began to be used in western Europe, printing was used for graphic portrayal; by 1472 it had been adapted to maps. Except for some private mapping, military sketching, sea charting, and so on, most maps thereafter were made to be reproduced.[343] From the late fifteenth until the beginning of the nineteenth century, there were two major methods of reproducing cartographer's drawings—woodcut and copperplate engraving.

Woodcut maps were used primarily in books, because the wood block could be locked into the form and printed along with the text on the same sheet of paper. For several centuries an enormous number of maps in the aggregate, largely views of cities and general maps, were reproduced this way, especially in cosmographies. The technique has significant limitations, however, since detail, small lettering, and subtle shading are all but impossible by the woodcut process.[344] Few if any thematic maps were made by woodcut until a variant of it, wood engraving, developed in the nineteenth century, when it was used for many small maps in books, magazines, and encyclopedias.[345] Relatively few thematic maps were made by wood engraving, however, and those that were were very simple (FIG. 80).

Until the latter part of the period when thematic mapping was developing, copperplate engraving was the usual method for reproducing thematic maps. Copperplate engraving was a most versatile technique, being capable of very fine as well as reasonably bold lines, small precise lettering, and, at least later in the period, rather effective shading. The major disadvantages of copperplate engraving were that it was expensive in material and labor, and that it was slow, both in the execution of the plate and in its printing.[346] A further drawback, which also added to its expense if a map were to accompany a text, was that a copperplate engraving could not be printed along with letterpress type, as could a woodcut or wood engraving; instead, the copperplate image had to be reproduced with a different kind of press.

The development of thematic cartography up to the early nineteenth century was not much retarded by the capabilities of the copperplate engraving techniques and craftsmen. Most of the subject matter and objectives could be effectively portrayed, and about the only hindrance was its cost. Had copperplate engraving been cheaper, there probably would have been more thematic maps before 1800, but innovative presentation and symbolism probably would not have occurred. Except for subtle shading, as for landform delineation, the requirements for new cartographic symbolism did not arise until the nineteenth century.

As I mentioned in chapter 3, the introduction of the new method of duplication, lithography, at the beginning of the nineteenth century turned out to be an important factor in the rapid development of thematic cartography. Although some maps (not thematic) were reproduced by lithography in the decade immediately after its discovery, that is, before 1810, it was not widely employed until after the publication of the English, French, and German editions of Senefelder's *Complete Course of Lithography* in 1818 and 1819.[347]

Although lithography differs fundamentally from the intaglio engraving process and the relief woodcut and wood engraving processes, it is not one single technique. The lithographic image is printed from an essentially flat surface, originally from smooth or grained (slightly roughened) limestone and later from grained metal plates, on which the oily printing and water-dampened nonprinting areas are kept separate because water and oil do not mix. The image of the printing surface can be obtained in two ways: it can be produced wrong-reading directly on the printing surface, or right-reading on some other material, which is then placed face down and transferred to the printing surface. Only after the period with which we are concerned was lithography combined with the photographic process.[348]

During the first half of the nineteenth century the majority of the maps reproduced by the lithographic process were probably prepared by lithographic engraving, a technique rather like copperplate engraving in that both involved working in reverse and cutting into a surface. Whereas the copperplate engraver literally incised the plate to make grooves to hold the ink to be printed, the lithographic engraver merely cut through a protective coating, called a ground, primarily to expose rather than to incise the underlying surface. When the lithographic engraving was oiled, the oil adhered only to the printing surface exposed by the grooves in the coating. After the ground was washed away, the image, when inked, would be ready to print. The two engraving techniques produced very similar graphic results, certainly in terms of general appearance and design.[349] Flat tones could be produced in a variety of ways, but by the 1820s this was accomplished mostly with line patterns made by ruling machines which could be used in either copperplate or lithographic engraving.[350]

The methods that could be used for producing tones or shading did differ in the two processes, but if competently done the results were not greatly different. There were, however, several differences between copperplate engraving and lithography that were important to thematic mapping. One was that lithography was considerably faster and therefore cheaper; consequently, many more thematic maps were probably made during the developmental period of the nineteenth century than would otherwise have been the case. Great expense is not conducive to experimentation, especially by individuals, and in a very real sense many of the thematic maps of this period were experimental.

A second difference is that very early in the lithographic era the process known as autographic transfer was developed.[351] In the transfer technique the image is prepared, right-reading, with a special ink on special paper and the drawing then is transferred upside down to the printing surface. It will then be wrong-reading, but when printed will be right-reading again. Although good-quality transfer materials and techniques did not come into wide use until after the mid-nineteenth century, the technique did allow the preparation of maps by less-than-skilled craftsmen who did not have the training and experience required to draw in reverse on a ground. This kind of autographic process probably resulted in a good many special kinds of thematic maps being prepared to illustrate reports, such as those on disease and sanitary conditions. Autography also allowed the relatively easy preparation of overlay material to be printed on already existing maps.

The development of lithographic techniques made color printing, rare before 1850, increasingly feasible. Printing solid colors over extended areas, impossible with the copperplate engraving process, was relatively simple in lithography.[352] In addition, the improvement of lithographic transfer techniques after 1850 made it easier to produce patterns and rulings. Color printing remained more expensive since each colored ink required a separate plate and printing operation, but the considerable problems of register (the printing of each successive color in its proper position with respect to the others) were gradually overcome.

For some time after color printing became relatively common (ca. 1850-60), color was used simply to do more quickly and cheaply what previously had been done by hand coloring. It was not until well after thematic cartography had essentially matured (by the 1860s) that color began to be widely used as an independent, representative element in thematic map design, which allowed more information to be displayed with clarity.[353]

It is generally thought that the proliferation of duplicating methods in the nineteenth century, especially lithography, had a great influence on the development of thematic cartography.[354] From the point of view of expense it undoubtedly did, simply by making more maps possible. On the other hand, it is likely that the burgeoning of thematic cartography had an equally great influence on the development of the duplicating processes used for maps. The need to portray classes and rankings on choropleth maps as well as quantitative variations within distributions of all kinds, from moral statistics to the land surface form, called for techniques for printing uniform (flat) and continuous (variable) tones, or at least reasonable approximations of them. Both copperplate engraving and lithography met the challenge, and in the hands of skilled craftsmen the two methods were equally effective up to about 1850. From then on, lithography clearly had the advantage.

### THE PRODUCTION OF TONE

Until the end of the eighteenth century the need was minimal in cartography for different flat tones or for smooth tonal gradations from light to dark. Occasionally a flat tone would be wanted to indicate water bodies, sandbars, shoal areas, and the like, but more often the oceans would be given an overall pattern of wavy or broken lines, or be sprinkled with drawings of ships or sea monsters on the more decorative maps. On many maps the coastline was edged with parallel dashed lines. When a flat tone was used it commonly served only to make distinctions between different kinds of areas, such as land and water or deep and shallow areas. Although the latter is obviously quantitatively based, the intent of the representation was usually far more qualitative than numerical in any statistical sense.

The need for qualitative distinctions among areas on both general and thematic maps could easily be met by the application of hand color, but obtaining flat tones on a monochromatic map was much more difficult. Mechanical ruling machines, devised for use in copperplate engraving at the end of the eighteenth century, made it possible to produce such tones by engraving line or stipple patterns either directly in the metal plate or in a wax ground (for later etching).[355] When the ruling machine was adapted to lithography in the early nineteenth century, the image was incised in the thin protective gum arabic ground and then treated to form a printing image by oiling the stone surface thus exposed. Ruling machines were ideal for cartographic work because the lines could easily be started and stopped at the edges of the area to be shaded. Darker or lighter tones could be obtained by varying the spacing and the thickness of the lines or by employing two sets of lines as in crosshatching.

The ruling machine was the most mechanical of the various methods of obtaining flat tones, but there were other ways it could be done in copperplate engraving. This is not the place to go into detail on these techniques, but a brief review is in order because they were often essential in thematic cartography for both qualitative and quantitative representation. The aim in all cases was to produce a series of closely spaced, tiny depressions in the copper. This could be done directly on the copper plate by numerous flicks with a sharp tool, by using a punch and hammer to make dots, or by using a variety of rough-surfaced rollers, rockers, and roulettes which would produce a pattern of indentations. More and deeper indentations produced a darker tone (FIG. 97). The same techniques could be used in a wax ground covering the plate, and then etching with acid would produce the depressions in the plate; within limits, the longer the etch, the darker the tone.

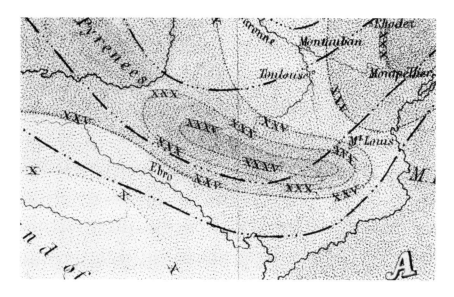

**Figure 97** Part of the 1848 "Hyetographic or Rain Map of Europe" in Johnston's *The Physical Atlas*, enlarged 100 percent, showing the progression of tones produced by increasing application of a roller in different directions on the copperplate. (Author's collection.)

Another method of obtaining a tone on a copperplate was by the aquatint technique, also an etching process but one in which the acid-resistant ground consisted of closely spaced, scattered specks of some compound.[356] The subsequent biting with acid of the metal surface left exposed by this open ground produced a fine-textured network or grain of irregular etched lines that could be inked and printed. A variety of grounds could be applied to the plate in different ways, such as a liquid which cracked or crazed when it dried, or the ground could be dusted on as a dry powder and affixed to the plate by heat. Each method produced a distinctive pattern and texture of grain. A kind of "masking" process used a varnish "stop" painted over areas to be protected from the acid, which stop later could be lifted or stripped away. An area bitten twice would have a darker tone than an area bitten once, and so on (FIG. 98).[357]

Lithographic methods for producing tones by techniques similar to aquatint were also developed.[358] They do not appear to have been as widely used for thematic maps, because in lithography the production of smooth tones by other means was considerably easier. In addition to the widely used machine ruling to obtain tones in lithographic engraving, shading could be done on a grained (slightly roughened) stone surface with a wax crayon. This would leave flecks of crayon on the tiny projections, which would then print a tone. As the lithographic techniques developed, crayon shading was done on grained (rough-surfaced) transfer paper. Other kinds of patterns, such as ruled lines or dots, were also transferred instead of being produced directly on the lithographic surface.[359]

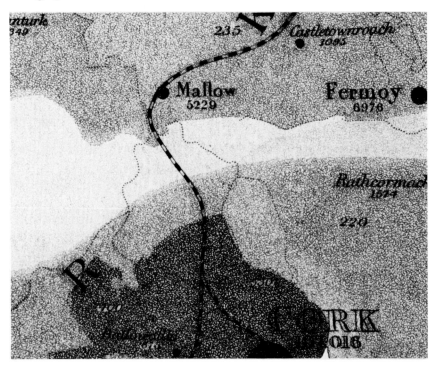

**Figure 98** Part of Harness's 1837 map of population in Ireland (FIG. 53), enlarged 100 percent, showing a progression of flat tones produced on the copperplate by aquatint. (Author's collection.)

The production of flat tones was necessary when the portrayal was choroplethic, that is, when statistical averages for enumeration districts were mapped. On the other hand, if the cartographer assumed that the data varied from place to place within the enumeration district, this called for the use of continuous tones from dark to light instead of flat tones. Smooth gradations were much more difficult to obtain.

In copperplate engraving continuous tone effects were added by aquatint (FIG. 99), by skillful use of rockers and roulettes, or by variable etching of mechanical ruling. After such a copperplate had been prepared it could be modified to obtain the desired continuous tone effect by burnishing down (lessening) the roughness of the plate as in the mezzotint process (FIG. 100).[360] Some map engravers were exceptionally skillful, for example, Dower, who prepared some of the maps designed by Petermann when the latter was in England (FIG. 56) and Olsen, the versatile Danish cartographer (or his engraver) (FIG. 24).

## New Techniques and Symbolism

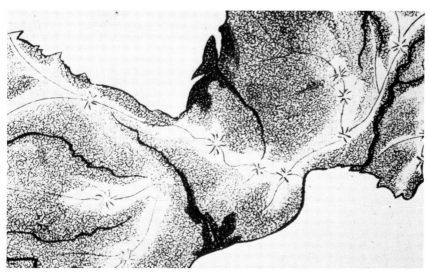

**Figure 99** Much enlarged section of Ritter's 1806 "...Europa als ein Bas Relief..." (FIG. 39) showing variable degrees of biting in copperplate aquatint. (Courtesy of the Royal Library, Copenhagen.)

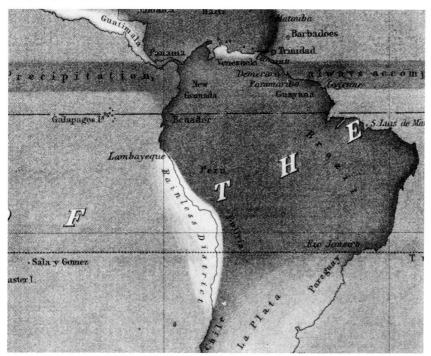

**Figure 100** Detail of the "rain map" in Johnston's *The Physical Atlas*, Meteorology, no. 4, showing burnishing on a roughened copperplate. (Courtesy of the British Library.)

The production of continuous tone was much easier in lithography than in copperplate engraving. It could be done by employing crayon shading, either directly on grained stone or, later, on grained transfer paper. This had the real advantage that, while doing the shading, one could see the effect. Although the printed result might be either darker or lighter overall, at least the tonal variations were directly visible, which was not the case with the continuous tone processes in copperplate engraving. Petermann prepared crayon-shaded population maps in this way, one of England and Wales (FIG. 58) and one of Scotland, to accompany the report of the 1851 census. A portion of the map of Scotland is shown full size in FIGURE 101.

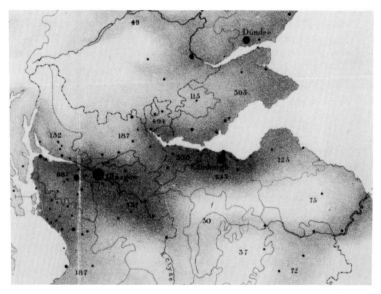

Figure 101 Part of Petermann's population map of Scotland in the *Census of Great Britain, 1851,* showing continuous tone by lithographic crayon shading. Full size. (The white streak is caused by a fold.) (Courtesy of the British Library.)

PORTRAYAL OF DENSITY AND QUANTITATIVE AREAL DATA

As mentioned earlier, one of the important tasks of the thematic cartographer is the portrayal of quantitative variations from place to place so that the structural characteristics of a distribution may be revealed. There are several ways to do this: by choropleth, by continuous-tone shading, and by what has come to be called the dot method. All of them are systems to show the variation of "numerousness" per unit area, or density.

Many earlier maps, which can hardly be called thematic, implied density of various phenomena by mapping the locations of the individual units in some class, such as mills or churches. Simply by their relative proximity to one another the mapping suggested density, but maps in which calculated density values were symbolized by various flat tones or patterns did not appear until systematic censuses gathered data by enumeration districts. So far as is now known, the first such map was the choropleth map of instruction in France made in 1826 and published in 1827 by Baron Charles Dupin (FIG. 75). Dupin's transformation of numerical, tabular data into a cartographic display apparently opened the eyes of a great many scholars with its visual impact. Although the well-known geographer Malte-Brun had coined the expressions *la France obscure* and *la France éclairée* four or five years earlier to characterize the unenlightened and enlightened sections of the country, Dupin's map was so effective that the use of those terms was often ascribed to him.[361]

Some of the first printed choropleth maps had no legends showing a scale of tones with assigned magnitudes. Instead, each numerical value was shown on the map and an attempt was made to give each district a tone graded according to its exact value. The engravers were unsuccessful in making such no-class maps, since they did not have full control over the darkness of very many flat tones.[362] So far as we now know, the first choropleth map to provide a legend and give class limits to the tones was the 1828 Prussian manuscript map of population density (FIG. 50). After the early 1830s most choropleth maps employed classes of various sorts, some based on rankings of the districts, some based on percentage departures from a mean, and some based on categorizing the data itself. The French cartographer Minard, as late as 1866, complained that Dupin's map (FIG. 75) and those of Balbi and Guerry (FIG. 76) were unsatisfactory because one could not relate properly the levels of shading to the numbers they represented. He devised a system which would make that possible, but apparently it was never used by anyone else.[363]

A refinement of the choropleth method is today called the dasymetric method, a name given to the technique only in 1923 by the Russian geographer-cartographer Semenov-Tian-Shansky. Although Scrope made a small map of world population density in 1833 which is technically the first dasymetric map, it is simply the shading of some highly generalized areas. It is much more appropriate to credit Harness's 1837 population map of Ireland as being the first (FIG. 53). Instead of merely using the boundaries of administrative units, the dasymetric method employs more meaningful divisions for the shaded numerical categories. The system is described in detail in chapter 5. Although Harness's sophisticated use of the dasymetric method was apparently not employed again until near the end of the nineteenth century, small, generalized maps of this class were not uncommon.

The primary purpose of the dasymetric method is to reduce the emphasis placed on administrative boundaries by the ordinary choropleth technique. Even before the dasymetric idea occurred to anyone, Quetelet in Belgium complained about Dupin's map of shaded *départements* (FIG. 75), and he devised an alternative way of portraying variations in ratios. As described in chapter 6, Quetelet's method was to use continuous tone—the greater the ratio, the darker the tone (FIG. 77). Quetelet's system of smooth tonal changes was not widely followed for population and man-related data, though Petermann used it for his cholera map of Britain (FIG. 91) and his population maps based on the 1841 and 1851 censuses (FIGS. 56, 58, and 101). This technique, apparently originated by Quetelet, must quickly have become well known throughout western Europe, since his maps, which originally appeared in 1831, were also included, unchanged, in his very influential *Sur l'homme et le développement de ses facultés; ou, Essai de physique sociale* (1835) which was translated into both English and German. The book created such a stir that it is unlikely that any educated person in western Europe was unaware of it. Quetelet's technique of continuous tonal shading, the more the darker, was adopted by Olsen for precipitation (FIG. 24) and by Berghaus in his *Physikalischer Atlas* for numerous maps of distributions, ranging from precipitation to zoological densities and earthquakes. Whether Petermann obtained the idea from Berghaus or directly from Quetelet's maps will probably never be known. By mid-century the continuous-tone technique was widely employed in thematic mapping, either by aquatint or by manually applied stipple in copperplate engraving or by crayon shading in lithography.

A curious aspect of the development of symbolism in thematic cartography is why the ordinary dot map, in which each dot represents a given number of items in some distribution, did not come into wide use at an early date. The dot map and the "darker the more" shading originated by Quetelet are conceptually very much alike, since in each case the darker areas show the regions of greater numerousness or density. Except for Frère de Montizon's 1830 dot map of the population of France (FIG. 49), true dot maps did not come into use until much later.[364] Berghaus came close in several maps in the *Physikalischer Atlas*, such as FIGURE 102, an inset map of Central and North America on his "Ethnographische Karte von Nordamerika" showing the distribution of Europeans and Africans, on which he portrayed the distribution of the "transplanted" Negro peoples by stippling. The last sentence of the legend reads in translation:

> The closer together the dots the denser is the population of the coloured.

Petermann also nearly used the dot map concept in a map of the popu-

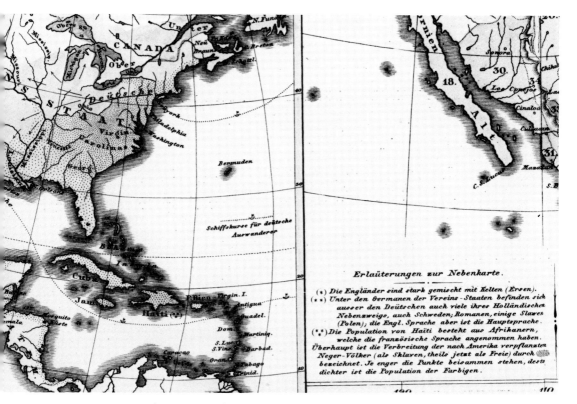

**Figure 102** Part of an inset map of the eighth Abteilung, Ethnography, no. 17 in Berghaus's *Physikalischer Atlas* showing by stippling the relative density of peoples from Africa. Full size. Copperplate engraving. (Courtesy of the Newberry Library, Chicago.)

lation distribution in Seibenbürgen in which he employed proportional circles scaled "as nearly as possible to the number of inhabitants of all the villages," but he did not use a strict unit value for the dots (FIG. 103).[365]

Apparently the first true dot map after the one by Montizon is one by a Swedish Army officer, Thure Alexander von Mentzer, dated 1859, showing population distribution in the Scandinavian peninsula.[366] An 1863 dot map showing the distribution of Maoris by means of crosses, with "approximately each cross representing about 100 souls," was appended to a report on roads and military settlements in the North Island of New Zealand,[367] but there seem to have been no others until near the end of the century. Why such an apparently obvious symbolic system did not develop while others conceptually far more complicated came into use much earlier is indeed a mystery.

The facts of density and other kinds of relationships can also be represented by isolines showing the ups and downs of a "statistical surface," and we will examine the development of that important technique later. First, I will trace the evolution of the use of proportional point and line symbols.

**Figure 103** Petermann's 1857 near dot map of the population of Siebenbürgen in *Petermanns Mitteilungen*. Lithograph. (Courtesy of the University of Wisconsin Geography Library.)

PORTRAYAL OF QUANTITATIVE POINT DATA

The representation of quantities at places has a curious history. The first two such quantities, depths of water and compass declinations, employed conceptually very abstract symbolism, and simpler representations of point data came much later.

Although the depth of water in dangerous areas must have concerned the earliest of mariners, apparently the practice of actually marking numbers on charts did not begin until the mid-sixteenth century.[368] By the first third of the seventeenth century charts began to show many soundings. The clarification of the profusion of numbers was soon accomplished by the use of the abstract isobath, a line identifying positions of equal depth. Soundings and isobaths continue to be standard features of charts today. Elevations on land and their corresponding symbolic delineation by the contour came later.

The only other quantity with locational significance deemed worth putting on maps seems to have been the values of compass declination. The earliest attempts to map such magnitudes also appear to have been in the early sixteenth century, and they too were soon delineated by abstract lines of equal declination. The development and application of lines of equal value in cartography will be described in the next section of this chapter, but their early use in connection with point data is noted here because no such sophisticated symbolism was applied to any other kinds of geographical quantities until very much later.

Ordinal distinctions of larger, smaller, and so on, had of course appeared on maps in medieval and perhaps even earlier times in the representation on general maps of more or less important cities and ports, the rankings of which probably depended in part on the sizes of their populations. These distinctions were represented by arbitrary geometric shapes, such as circles, by perspective miniatures of groups of buildings, and by plan views, all graded according to size.[369]

The use on maps of what are now called point symbols to represent stated quantities at places, rather than mere ranks, is a nineteenth-century development. Their introduction was slow. Apparently the first map to show numerical population data is plate 6 in Ritter's atlas, *Sechs Karten von Europa,* but it does so only by writing numbers in appropriate places. In 1815 J. Wyld prepared a curious map of the world on which he showed populations by placing numbers in each nation, appending to the numbers the two signs -M and -T to show that the numbers referred to millions and thousands (FIG. 104).[370]

Wyld used two other kinds of proportional symbols which can be called point symbols by stretching things a bit. One of his objectives was to show the religious composition of delimited areas, not necessarily nations, and he did this by outlining each area with a colored band, sections of which were of different hues. He expressed his symbolism as follows:

> The prevailing Religion being indicated by the length
> of the line in color; other sects less numerous than the
> first are marked by shorter lines of color.

The colors were keyed to a legend that listed the religions and tallied their adherents: thus the area labeled "Germans" is outlined and bounded by a band consisting of a length of red (Catholics), of green (Protestants), of black (Jews), as well as one of brown (absence of religion). Needless to say, the occurrence of the various religions is not difficult to note, but the relative proportions are much harder to perceive. Today this would be done by means of sectored circles.

Wyld's third objective, to show degrees of civilization, was symbolized by roman numerals ranging from I = savages to V = most civilized. The numerals were places in the national areas and were varied in size to indicate degrees of prevalence. For example, in the area labeled "Americans" (which encompassed only the region between the Rocky Mountains and the eastern coast) there is a small I and II with a somewhat larger IV. No part rated a V. The use of varied sizes of roman numerals is probably the first bona fide attempt in thematic cartography to employ proportional or graduated point symbols to portray magnitudes, in this case ordinal rankings.

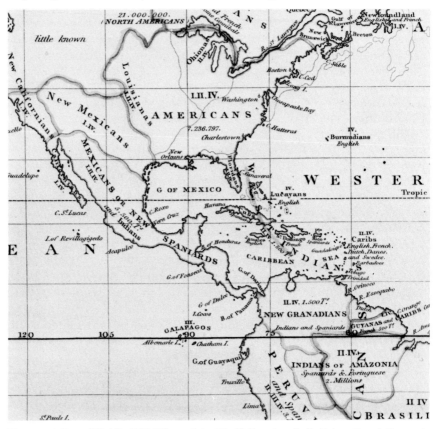

**Figure 104** Part of Wyld's 1815 "Chart of the World Shewing the Religion, Population and Civilization of Each Country." Reduced 25 percent. Copperplate engraving, hand colored. (Courtesy of the British Library.)

The use of proportional point symbols on maps to represent magnitudes more informative than ordinal ranking, namely actual quantities, came much later. The first to do so appears to have been Harness on his 1837 population map of Ireland (FIG. 53). He stated his intentions as follows:[371]

> The towns are represented by dark spots, of which the areas are regulated by the number of their inhabitants.

By that he meant he had scaled the sizes of the circles so that their areas were strictly proportional to the populations of the towns they represented, easily accomplished by employing a constant scale for the diameters related to the square roots of their populations. Although Harness's intentions were clear and precise, their execution by the engraver introduced minor variations in the proper sizes of the circles, probably unavoidable in the technique, especially in the smaller sizes. Nevertheless, Harness's use of proportional circles for urban populations on maps was innovative.

The next use of proportional circles on a map seems to have been by Minard in 1852 to show the tonnage of the shipping in the various ports of France.[372] He also included a legend consisting of a series of circles with the corresponding numerical tonnages.

Petermann also employed proportional circles in this same period, the first time on a map of the distribution of the population of the British Isles prepared in 1851 for the National Society (FIG. 57). Although drawn by Petermann, the map is identified as designed by Samuel Clark. It is not at all unlikely that the use of proportional circles was the result of Harness's population map of 1837.[373] We do know that Petermann was acquainted with Rawson, who reviewed the Railway Commission report for the Royal Society, since they both served on the committee for Section F (Statistics) for the nineteenth meeting of the British Association at Birmingham in 1849. Although Petermann certainly did not originate the practice of using proportional circles to represent magnitudes at places, he early realized their utility as a cartographic symbol and used them extensively after 1851, always being careful to use "spots, the areal size of which corresponds as nearly as possible with the number of inhabitants."

In addition to the shaded population distribution maps by Petermann (FIGS. 56 and 58), the census of 1851 of Great Britain included two maps showing administrative divisions and registration districts by W. Bone.[374] The two maps by Bone contain legends showing circles for seven sizes of towns and include the statement:

> Towns of Population ranging between the above
> numbers are indicated by dots of intermediate size,
> proportioned to the respective populations. The dots
> are not intended to represent the area covered by the
> Towns but have reference entirely to Population.

Presumably the 1852 maps by Minard and Bone are the earliest proportional symbol maps to contain legends showing a representative set of circles.

Before proportional circles were used on maps they had been employed much earlier, both on the Continent and in Britain. The graphic portrayal of numerical data is at least as old as the tenth century.[375] By the mid-eighteenth century, means to represent geographical statistics were certainly available, even though such data were rare. A brief review of the noncartographic use of such symbols as proportional circles and squares is in order because their use on maps probably stemmed from it. Three names stand out: A. F. W. Crome in Germany, William Playfair in England, and D. F. Donnant in France.

Crome, the author of the earlier noted 1782 map of Europe showing products (FIG. 16), was a teacher of geography, history, and statistics at Dessau and Giessen who was much traveled and became well known.[376] An early publication in 1785 includes a chart showing the relation among the sizes and populations of all the European states.[377] At least two other graphic displays by Crome are known, but they have neither dates nor places of publication.[378] One is concerned primarily with the areas of states in Europe shown by superimposed squares, but the other symbolizes population density with proportional circles in an unusual way, namely by scaling the circles on the basis of the number of units of area per person, so that the smaller the circle the greater the density. Crome was an active teacher and publisher from the 1770s until well into the first half of the nineteenth century.

William Playfair was the younger brother of John Playfair, the geologist-mathematician, and he tried his versatile hand at many things, ranging from being a draftsman with James Watt to journalism.[379] In 1786 he published his first attempt at statistical graphics, a volume containing graphs and charts concerning trade, imports, exports, revenues, and debts.[380] Although he called it *The Commercial and Political Atlas*, it contained neither maps nor the symbols that were to be commonly utilized on maps. It was translated into French in 1789, and it was apparently very well received. In 1801 Playfair published a small volume in which he employed proportional circles to represent the populations and areas of various cities and countries on four charts.[381] On one of the charts showing the magnitudes of the populations and resources of the countries of Europe, he sectored the circle representing the Turkish Empire to show its relative areas in Europe, Africa, and Asia. That was apparently the first use of a sectored, proportional circle which, surprisingly, did not show up on thematic maps until a half-century later.

That simple, rather incidental use of the sectored, proportional circle was soon applied, again by Playfair, in an addition to a translation of a statistical description of the United States by Donnant of France.[382] On the frontispiece, which Playfair added to the translation, he stated in the title that "This Newly invented Method is intended to shew the Proportions between the divisions in a Striking Manner." There may be some question whether Crome or Playfair first employed the proportional circle, but as of now it seems likely that Playfair was the originator of the sectored circle.[383]

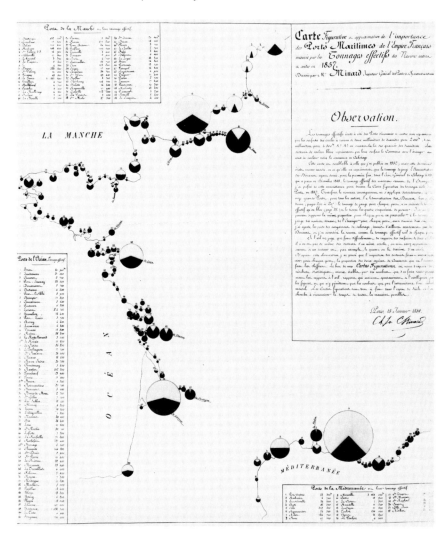

**Figure 105** Minard's 1859 map of tonnage handled by French ports. Original 550 × 630 mm. Lithograph, hand colored. (Courtesy of l'Ecole Nationale des Ponts et Chaussées, Paris.)

The first cartographic application of the sectored proportional circle appears to have been by Minard in a map dated 1858 showing the quantities of the various kinds of butcher's meat dispatched to Paris (FIG. 68). The following year he made another map showing the relative importance of the ports of France by graduated circles sectored to show the proportions of foreign trade and coastwise trade by coloring the former blue and the latter black (FIG. 105).[384] In the "Observation" on the map he pointed out that data distinguishing between foreign and coastwise trade had become available only in 1858. He also observed that, although the eye has difficulty in comparing the areas of proportional circles, it could easily judge the comparative sectors. Minard specifically disclaims being the inventor of the sectored circle by observing that "sections have already been employed before I used them,"[385] but whether he was referring to the sectored circle as used by Playfair or to some other use on maps we do not know.

PORTRAYAL OF QUANTITATIVE LINEAR DATA

It is not known when the humid-land reality of the increasing volume of rivers downstream became regularly represented by a thickening line on maps.[386] Early symbolism was quite arbitrary about the width of rivers on the figurative maps of the Middle Ages,[387] but as mapmaking moved toward more realistic portrayal of geographical phenomena, it was only natural to represent increasing actual width with a line of the same character. Furthermore, the natural association between increasing width of rivers and increasing volume of water must have been noted. It was then but a short step for J. G. Lehmann, at the end of the eighteenth century, to employ lines of varying width, called hachures, to show greater and lesser angles of slope.[388] That seems to have been the first use of a line of varying width to portray actual quantitative data as compared to the more intuitive use of the tapering line for streams. In the years following Lehmann's proposal, several variations were suggested by others, and the concept of the hachure as a kind of vectorlike symbol showing both direction and magnitude must have become very familiar, particularly to the military. Hachures had soon become a standard way of representing the land surface form on detailed maps, most of which were prepared by army organizations.[389] It is no surprise, therefore, that the first use of a proportional line to show linear, quantitative data on a thematic map was by an army officer, Lieutenant Harness, who in 1837 prepared the two flow maps described in chapter 5. From the description by Harness of the preparation of the maps, it is evident that he intended the widths of the flow lines to be strictly proportional to the numbers they were to represent, and one can only assume that the original drawing was correct.[390] Unfortunately, the engraver chose to use closely spaced parallel lines in straight segments to produce the bands of gray on the black-and-white map (FIG. 106). Although the overall effect is not greatly affected, such a technique is difficult to control, and the widths of the lines depart considerably here and there from strict proportionality. Perhaps because of the likelihood of this happening, or simply because Harness was a careful engineer, he showed all the numerical data on the map adjacent to the lines.

Flow lines were soon after employed by two other engineers, Belpaire of the Ponts et Chaussées in Belgium and Minard of the Ponts et

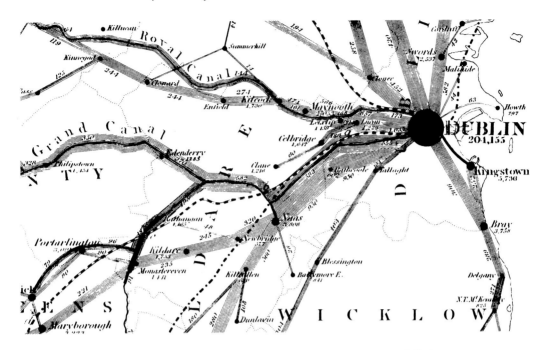

**Figure 106** Part of Harness's 1837 map of passenger movement by conveyance. Full size. Copperplate engraving. (Author's collection.)

Chaussées in France, to portray the movement of goods and passengers on roads, railways, rivers, and canals. Their delineation was more precise than that of Harness: they simply drew parallel lines the requisite distance apart and then colored the spaces between the lines to show the class or classes of data being portrayed.

Flow lines portraying quantitative data seem not to have been much used by geographers and students of the physical world. Largely because of Minard's long series of *cartes figuratives*, the flow line technique of portraying quantitative data soon become well known among statisticians and bureaus dealing with matters of commerce, especially in France.[391] Almost anything that moved could be mapped with flow lines. Between 1845 and 1869 Minard himself employed them to show flows of travelers, all kinds of merchandise on France's transportation facilities, international trade in coal, cotton, and wine, and even the flow of emigrants.

### Lines of Equal Value: Isolines

As may be apparent from the foregoing, most ways of representing area, point, and linear quantitative geographical data on thematic maps had their origins in qualitative representation on general maps. The reverse is the case, however, with the line of equal value.

Isolines came into use specifically to portray quantitative data, and with few exceptions their use was largely thematic until the nineteenth century. Before taking a brief look at their development I should emphasize that the lines can be used for very different purposes.

Although a line of equal value is merely a simple mark on a map, not even having the visual complications of the line of proportional value, the graphic simplicity of the isoline is deceptive, for it can be used to portray concrete and abstract phenomena that have not the slightest conceptual relation with one another. The line of equal declination, called "tractus chalyboclitici" by Kircher and "curve lines" by Halley, portrays an extremely complex relationship between a magnetic force field and an arbitrary directional system at the earth's surface. The contour is completely different in concept. It is the trace of a horizontal surface with some real or abstract, continuous, statistical surface reckoned from some datum and then projected orthogonally to a map, a much simpler geometric concept. Again, quite different are lines showing simultaneous blooming times for some specific plant or the arrival times of some species of migratory bird, values which have no geometric foundation at all. So far as is known, no mapmakers who employed lines of equal value commented on any conceptual differences or similarities among the kinds of data being portrayed by them until Humboldt and Berghaus related the isotherm to Halley's curve lines.[392]

The history of the development of the line of equal value as a cartographic symbol is much more complicated than that of the other systems of representation in thematic mapmaking, and it is too involved to be told in detail here.[393] Nevertheless, a summary is appropriate, for this symbol is one of the most used in thematic maps, and perhaps more than any other it epitomizes the fundamental aim of portraying the structure of a geographical distribution. I will begin with the isobath contour, perhaps the simplest of the ways this kind of line is employed. In the following I will use the term "contour" to refer to any line showing actual or assumed elevation above or below a reference surface or datum.

### Contours and Layer Tinting

The contour had tentative beginnings as early as the end of the sixteenth century in the form of lines of equal depth on a map of the River Spaarne made in 1584 by the Dutch surveyor Pieter Bruinsz (FIG. 107).[394] The next use of this symbol came more than a hundred years later in 1697 by Pierre Ancellin, again in the Low Countries.[395] Maps by Blackmore in 1715,[396] by Cruquius in 1730, and by Buache in 1737 were finally followed by Buache's printed map of 1752 showing depths in the Channel, which quickly became famous (FIG. 35).[397] In 1782 Dupain-

*New Techniques and Symbolism* 211

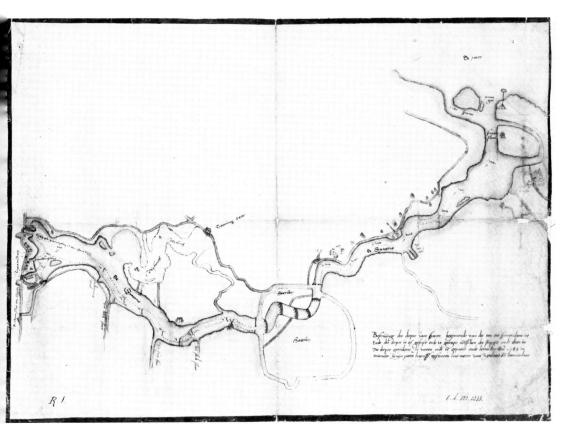

**Figure 107** Bruinsz's 1584 map of Het Spaarne. Original 630 × 460 mm. Manuscript, paper, colored pen drawing. (Courtesy of Hoogheemraadschap van Rijnland, Leiden.)

Triel published a small volume devoted to a proposal by du Carla, who had earlier suggested using contours to show the configuration of the land surface. The publication included a hypothetical map illustrative of the system (FIG. 108).[398] Du Carla's suggestion and Dupain-Triel's two maps (FIGS. 38 and 109) made a profound impression, being frequently cited in the first part of the nineteenth century and even as late as nearly midcentury in the *Physikalischer Atlas* and *The Physical Atlas*. It was not until the late 1820s, however, that enough elevational data became available for Olsen to prepare a thematic contour map of Europe (see chap. 4) which was later copied by both Berghaus and Johnston.[399] By the 1840s the contour was well known and widely used.

Contours by themselves on small-scale maps are not very expressive of the land surface form unless the distance between contours is relatively small and the map reader has learned to visualize the three-dimensional surfaces from which they are derived. It is not surprising, therefore, that attempts were made to enhance the graphic quality of the small-scale maps soon after contours began to be employed. The portrayal of the form of the land surface has been a perennial cartographic challenge and has given rise to a great many techniques. Its history is exceedingly complex, with a voluminous literature; it is far too large a topic to be treated in this book. Nevertheless, a brief review of one aspect is appropriate, because it is closely allied to thematic map-

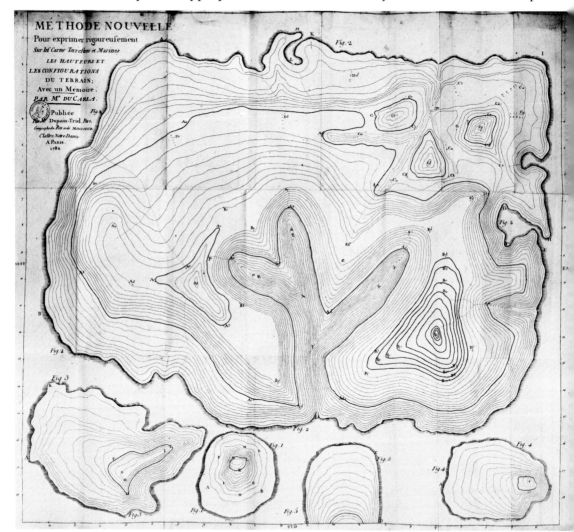

**Figure 108** Du Carla's 1782 "Méthode nouvelle pour exprimer rigoureusement sur les cartes terrestres et marines les hauteurs et les configurations du terrain." Original 650 × 516 mm. Copperplate engraving. (Photo. Bibliothèque Nationale, Paris.)

## New Techniques and Symbolism

ping—namely, the attempt to enhance the graphic quality of a contour type of map by employing colors or various uniform tints as area symbols between selected contours, a technique commonly known as layer tinting.

The first layer-tint map seems to be the second contour map of France made by Dupain-Triel in 1798–99 (year VII) (FIG. 109). He used the contours of his earlier 1791 map (FIG. 38), but to these he added sinuous ridge lines and at least four, if not more, tints between the contours. It is not easy to discern how the tints were applied to that prototype, but however it was done they were not well controlled, so that the configuration of France leaves much to be desired. The next layer-tint map, also a prototype, is apparently one made by the Swedish botanist G. Wahlenberg in 1813 to accompany his description of the flora of the Tatra mountains.[400] Although Wahlenberg's "contours" are not exact, they clearly bound four elevational zones. What is new is that he employed different colors to portray the "layers."

The next layer-tint map, also still a prototype, is the manuscript map of 1824 by Olsen and Bredsdorff made for the Mountains of Europe competition sponsored by the *Société de Géographie de Paris* referred to in chapter 4. Their map utilizes uniformly spaced hachures of different width in a graded series to show elevation (FIG. 110). It is obvious that they intended it to be a layer-tint portrayal, but it was not very successful. Steinhauser reports that Ritter von Hauslab, a strong proponent of contours, colored depths on a chart with lighter to darker blues in 1830.[401] Also in 1830, the Swedish cartographer C. Forsell prepared a map of the southern part of the Scandinavian Peninsula in which he employed green, red, yellow, and white for the altitude zones: 0–300, 300–800, 800–2,000, and over 2,000 feet.[402] E. von Sydow began using layer tints in 1837 in his physical wall maps of the continents, which enjoyed great popularity for many years (see FIG. 42).[403]

The versatile and imaginative Larcom appears to have produced the first layer-tint map in Britain. For an 1845 report on the "Occupation of Land" in Ireland, Larcom prepared a map at a scale of 1:633,600 described in the report as follows:[404]

> A Map of Ireland, prepared at the Ordnance Survey Department, by Capt. Larcom, R. E., accompanies the Index. This map shows the several localities visited by the Commissioners.... It likewise gives the classification of the country as regards its altitude—every portion of the land having been brought within one of five different zones... (0–250, 250–500, 500–1000, 1000–2000, over 2000 feet).

The layer tinting was applied by watercolor, in the order from the lowest to the highest: yellow, light red, dark red, brown, and black.

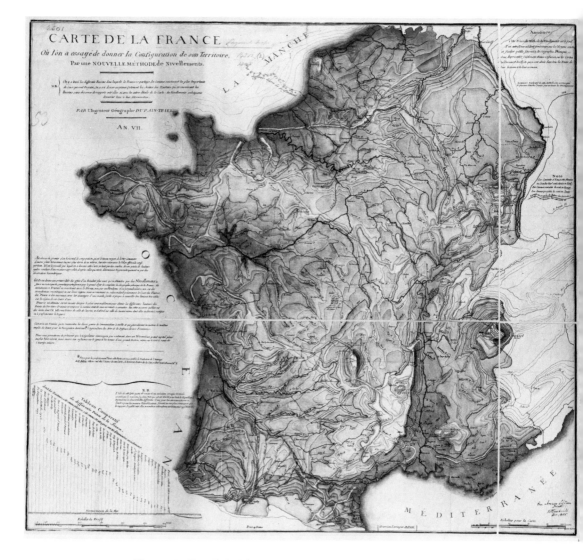

**Figure 109** Dupain-Triel's 1798–99 (year VII) map of France with layer tints. Original 530 × 482 mm. Copperplate engraving, method of shading in doubt. (Courtesy of the British Library.)

The application of "layer tinting" to meteorological maps had already been done by Berghaus on his map of precipitation in Europe for the *Physikalischer Atlas* in 1841, and after the 1840s the technique was commonly used on many thematic maps to enhance the visual quality of quantitative data portrayed by lines of equal value.

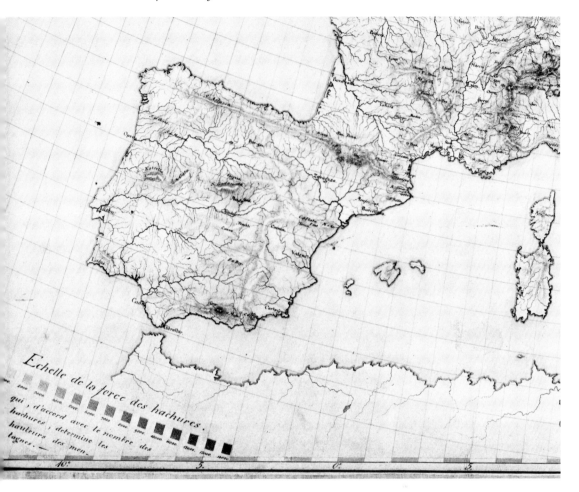

**Figure 110** Map by Olsen and Bredsdorff (1824) prepared for the Mountains of Europe competition of the Société de Géographie de Paris. Original 866 × 630 mm. Manuscript, paper. (Photo. Bibliothèque Nationale, Paris.)

### Curve Lines and Isopleths

Another early application of the lines of equal value was to portray the pattern of compass declination, a use which dates back even to before the time depth contours were employed. It appears that as early as 1536 the Spaniard Alonzo de Santa Cruz thought that lines of equal declination paralleled meridians. Several other attempts to map the declination are known, as noted in chapter 4, but it was Edmond Halley and his published maps of curve lines that made known generally the use of lines of equal value to portray the variations of a field force. Halley's curve lines of equal declination were followed in 1721 by

Whiston's lines of equal dip and by Humboldt's 1804 map of isodynamic zones.[405] That may have been the first time the prefix "iso-" was attached to a descriptive cartographic term, with which Humboldt later also coupled "-therm" and from which spawned well over a hundred iso-terms introduced since that time.[406]

Humboldt's application of the line of equal value to the portrayal of meteorological data was soon taken up by others. Kämtz used lines of equal value to map mean variation of the barometer in 1827, and Berghaus used them to map barometric pressure in 1838, precipitation amounts in 1841, and the number of stormy days in 1842.[407] After the mid-nineteenth century such portrayals were commonplace on thematic maps of geophysical fields.

The use of lines of equal value in biological mapping did not occur as early, nor was it as common, as it was in the geophysical sciences. Quetelet in 1842 proposed mapping lines of simultaneous blooming time for specific plants, and in 1855 Middendorf proposed using lines of average concurrent arrivals of particular kinds of migratory birds.[408] Other uses of these kinds of isolines came much later.

It will have been noted that the kinds of data being portrayed by these early uses of lines of equal value were obtained by observations made at specific geographical locations. It made no difference whether the measurements had to do with a magnetic field, an actual land or sea bottom surface, a static pressure or temperature field, or even with some phenological phenomenon, such as blooming dates. In each instance the collected data had specific locations from which a total distribution could be generalized and portrayed. To complete the gamut in thematic cartography, it only remained for the line of equal value to be applied in a more abstract manner to map data which did not have specific locations. This was first proposed in 1845 by the French engineer-mathematician Léon Lalanne.[409]

Lalanne was an engineer in the Ponts et Chaussées who had become interested in graphic calculation and representation, about which he presented a *mémoire* to the Académie in 1843. He also had prepared a considerable appendix, "Sur la représentation graphique des tableaux météorologiques et des lois naturelles en général," for the French translation of Kämtz' well-known treatise on meteorology.[410] In that appendix Lalanne provided graphic representations of more than forty numerical tables, to some of which he applied the "ideas of du Carla and M. de Humboldt" to show $z$ dimensions on an $xy$ plane.[411] The extension of the idea to a real map was not difficult, but in this case Lalanne had in mind specific population, that is, density, an abstraction in which the derived values referred not to particular places but to the whole areas of each enumeration district. In commenting on a suggestion by Morlet regarding *courbes d'égale excentricité* (of population), Lalanne combined the ideas of Humboldt and du Carla as follows:[412]

> It is again an additional application of the happy and fertile idea of which M. de Humboldt has given a celebrated example in the construction of isothermic curves. But this is not the only application possible, or the most important, having to do with the elements of the population.
>
> Suppose, in effect, that one partitions the territory of a country into a large number of sufficiently small parts so that it would provide a division as extensive as the communes of France; that at the centre of each of these divisions one raises a vertical, proportional to the specific population, or in other words, to the number of inhabitants per square kilometre in the territory of the commune in question; that one joins the extremities of all these verticals with a continuously curved surface, and finally one projects on a map, at a convenient scale, the contours traced on that surface which correspond to equidistant integral elevations: One will thus have lines of equal specific population and one will be able to observe the series of points along which the population is 30, 40, 50, ..., 100 inhabitants per square kilometre. A map of this kind will afford immediately a representation of the distribution of the population as exact and expressive as possible. Like a topographic map made according to the principles of du Carla, it would show some undulations, steep summits, craters, passes and valleys. One could conjecture that its relief might appear as the inverse of that of the surface of the ground; thus our populous valleys would appear as chains of mountains while, conversely, the deserted summits of our mountains would look like deep funnels.

Lalanne never made the kind of map he visualized, but the Danish cartographer, Ravn, did (FIG. 59). Two such maps appeared in 1857, probably on the suggestion of the engineer and mathematician C. Andrae, the author-inventor of the electoral system of proportional representation.[413] They were prepared exactly as Lalanne had suggested.

The previous year, in 1856, Petermann published a series of maps to accompany his study of the distribution of the most important agricultural products of the United States.[414] Ten of the fifteen maps show quantitative distributions in essentially the same way as Ravn's maps do. Petermann refers to the divisions between classes as "natural boundaries," but it is doubtful that he conceived of the quantitative distributions as being three-dimensional, as had Lalanne and Ravn.[415] Ravn's lines were soon named "isopleths" by von Sydow. In a long article on the state of mapping in 1858 he discussed Ravn's maps and wrote:[416]

> The author [Ravn] properly equates the curves of his map with equidistant height curves, and as these are lines of equal height or "isohypses," so might we suggest to name the lines of equal population quantity "isopleths" (from the Greek *isos* plus *plethos*).

The isopleth provides an impressive example of the cumulative and international character of science, ideas, and thematic mapping. One of its early ancestors, the depth contour, first appeared in the Low Countries, and another, the magnetic curve line, in Spain, Italy, and England. The one immediate, more exact, parent of the isopleth, the contour on the land, was a direct descendent of the depth contour, while the other, more abstract parent, the isotherm, stemmed from the family of magnetic field curve lines. By the mid-nineteenth century the portrayal of the continuous land surface by contours had become common, and a variety of facts, such as temperatures and population densities, were sufficiently available so that mapping techniques and geographical data could be combined in a wholly new method of visualization and portrayal.

An isopleth map of population densities employs an involved, graphic, geometric symbolism for describing a three-dimensional surface to show the structure of an imagined "statistical surface" formed by the variations in ratios of people to areas. An "ordinary" contour map is in reality a very complicated system of representation, and the concept of a statistical surface of population densities is exceedingly abstract. That the two could be combined by the 1850s, and readily accepted, shows how far thematic mapping had come.

### Epilogue

At the beginning of this book the long history of cartography was likened to a succession of gently sloping plains of different elevation separated by more abrupt escarpments. The scarps represent times of rapid change, and they lie between periods of relative stability. As one moves across that analogical terrain toward the present, in general it is an uphill climb. Some of the escarpments are quite steep, representing a time when several major changes took place, as in the fifteenth century when Ptolemy's *Geography* and the new techniques of printing letterpress and graphics had profound effects on the field of mapping. I noted that a similar, major revolution in cartography occurred between the mid-seventeenth and the mid-nineteenth centuries, the time when thematic mapping developed.

A closer look at the scarp representing the thematic revolution reveals that it has a distinctive profile. It begins with a very gentle slope in the late seventeenth century, rises steadily but slowly during the eighteenth, steepens abruptly during the first half of the nineteenth century, and then slackens off thereafter to a less steep but steady growth to the present. The sharp rise between 1800 and 1860 represents the combined effects of greatly increased curiosity about the earth and man and numerous innovations in the techniques and conceptual aspects of mapping. This was the relatively short period when almost all the symbolism used in thematic mapping today was devised and put to use.

Although the profile is distinctive, it is not unusual. The gentle slope, the steep rise, followed by a flattening of the slope is the shape of what is called a logistic or sigmoid curve on a graph with rate of change for an ordinate and time for an abscissa. Like an elongated S pushed forward, a logistic curve seems to describe the growth patterns of a variety of human phenomena, from manufacturing production to the development of scientific fields. It would be surprising indeed if the course of thematic mapping did not fit the pattern, since after all it was a response to a surge of curiosity and intellectual development.

In these days of modern cartography, with its advantages of computers, plastics, remote sensing, instant communication, and so on, it is nevertheless worthwhile to look back. Often the basic truths and concepts of a field are made clearer by observing how the innovaters coped with problems no one else had ever faced. The solutions by Berghaus, Dupin, Halley, Harness, Humboldt, Lalanne, Minard, Olsen, Petermann, Quetelet, Ravn, and a host of others are standard procedure today. They give thematic mapping a grand heritage.

# NOTES

1. Reports concerning the map competence of modern primitives (such as Daly 1879, p. 2) are sufficiently common to inspire some confidence, but one must not jump to conclusions (Bagrow and Skelton 1964, pp. 19–28). An old but extensive survey is Adler (1910) as reported by De Hutorowicz (1911).

2. Maps have been described as the oldest of the graphic arts (Greenhood 1944, p. 1). For a comparison of map and language as media of communication, see Robinson and Petchenik (1976), pp. 43–67.

3. Bagrow-Skelton (1964), pp. 31–32, and Hsu, M.-L. (1978).

4. Skelton (1972), pp. 26–35.

5. Brown (1949), p. 6.

6. Skelton (1965), pp. 1–28.

7. The literature in the history of cartography is surprisingly extensive. A few general surveys and useful reference works are: Bagrow and Skelton (1964), Brown (1949), Crone (1966), Ristow (1973), Thrower (1972), and Tooley (1952).

8. Thrower (1972), pp. 10–16; Ungar (1935, 1937).

9. Bagrow (1943).

10. Bagrow and Skelton (1964), pp. 37–38.

11. Bagrow and Skelton (1964), pp. 148–49.

12. Howse and Sanderson (1973).

13. Brown (1949), pp. 12–57. Several examples appear in Daly (1879), and a detailed example is Myres (1896).

14. Bagrow and Skelton (1964), pp. 41–50; International Geographic Union (1964).

15. The contribution of the church to the character of medieval *mappae mundi* is well documented in standard histories of cartography, such as Brown (1949), pp. 81–112. The relation of maps with other graphic expression in art and architecture is less well known, and the author is indebted to Jeannine Schonta for her preliminary investigations. See Fleming (1955).

16. Skelton (1972), pp. 35–37.

17. Ramsay (1972).

18. Beste (1578), First voyage 1576, pp. 37–38.

19. Compare Skelton (1972), pp. 1–6.

20. Robinson (1976a).

21. See Petchenik (1979) for a thorough analysis of the fundamental distinction between the two classes of maps. See Robinson and Petchenik (1976), pp. 108–23, for a fuller discussion of the concept of structure in thematic mapping.

22. The earliest such "hybrid" now known, primarily a general map, seems to be the Han military map, dated pre-168 B.C. See Hsu (1978), pp. 52–54. Nothing so early is known in the West.

23. Brown (1949), pp. 208–40; Robinson (1973).

24. In old French measure one *toise* = ca. 1.95 meters or 6.4 feet, and one *ligne* = ca. 0.226 centimeter or 0.089 inch.

25. Howse and Sanderson (1973), pp. 79, 101; Ritchie (1973), p. 12.

26. Brown (1949), pp. 241–75.

27. Milham (1945), pp. 260–68; Robinson (1973).

28. Robinson (1975).

29. Woodward (1975b).

30. Verner (1975).

31. Verner (1975); Skelton (1958).

32. Ristow (1975).

33. Pearson (1978).

34. Harris (1975), p. 113.

35. One of the most fundamental technical developments in cartographic history was the introduction of photography, both for map duplication and later for aerial survey, but it did not have any great effect on thematic cartography except perhaps on cost and some aspects of design. See Robinson (1975), pp. 21–22; Koeman (1975).

36. Ivins (1953), p. 8.

37. Dingle (1972), p. 28.

38. Adams (1938), pp. 199–208.

39. Geike (1905), pp. 201–36.

40. Merz (1896), pp. 200–275.

41. Cannon (1978), p. 105.

42. Dingle (1972), p. 28 and *Dictionary of National Biography*.

43. Büsching (1762), pp. xvii–xxix.

44. Much of this section is based upon the excellent survey by Strasser (1975).

45. Also known as Prieur de la Côte d'Or (1763–1832).

46. Merz (1903), p. 562.

47. Peterson (1979), pp. 135–41.

48. John (1884), 1:88 ff.

49. *Théorie analytique des probabilités*, 1812; *Essai philosophique sur les probabilités*, 1814.

50. The name of the prolific Quetelet appears both with and without the accent, more often without.

51. Wright (1955).

52. McKie (1972b).

53. For example: Russia 1725; Denmark 1742; United States 1743; Netherlands 1752; Italy 1757; Spain 1774; Switzerland 1776; Portugal 1779; Czechoslovakia 1784.

54. The Statistical Society of London, founded in 1834, grew out of the third

meeting of the British Association, which had added a Statistical Section to its list of five. It stemmed from the desires of Malthus, Babbage, Jones, and Quetelet. The last had attended the British Association meeting. The prospectus contained four classes of statistics of concern to the society: economic, political, medical, and moral and intellectual. More than one hundred papers and reports were presented in the society's publications in its first ten years. Royal Statistical Society (1934). For a good summary of intellectual aspects during this period see Peterson (1979), pp. 1–20.

55. Merz (1896), pp. 175–81, 212–15.
56. Armytage (1965), pp. 66–79.
57. McKie (1972a).
58. Robinson and Wallis (1967); see also Crosland (1967), pp. 147–231.
59. *Encyclopaedia Britannica,* fifteenth edition (1978), "Encyclopaedia."
60. Thomas Burnet: *The Sacred Theory of the Earth: Containing an Account of the Original of the Earth and of all the General Changes which it Hath Already or is to Undergo, Til the Consummation of all Things* (1681); G.-L. Leclerc, Comte de Buffon: *Histoire naturelle, générale et particulière,* 44 vols. (1749–1804).
61. Finke (1792–93).
62. Jarcho (1970b).
63. E. C. Smith (1972), pp. 95–100.
64. E. C. Smith (1972), p. 104.
65. Dainville (1968).
66. G. Engelmann (1965), p. 233, reports it from Oehme (1961), who in turn cites it from a 1724 study of mapping.
67. Torrey (1938).
68. Adams (1938), pp. 433–39.
69. Hellmann (1909). The description in Latin is quoted on p. 9.
70. Hellmann (1909).
71. Stommel (1950) discusses both Kircher's and Happel's maps.
72. Taylor (1956), p. 239.
73. Halley (1686). The map has been often reproduced; see Chapman (1941), Hellmann (1897), and Thrower (1969a,b, 1978).
74. Thrower (1969a,b, 1978).
75. Thrower (1969a,b, 1978) gives a far more complete description of Halley's geographical and cartographic accomplishments.
76. Dainville (1964), p. 95.
77. Thrower (1969b), pp. 669–73.
78. Robinson (1973).
79. For example, it was normal practice to "run down a parallel" to reach a port, and if one did not allow for declination, when it was overcast or dark one might not only miss the port but run into something less desirable.
80. Hellmann (1909), pp. 10–11. Hellmann includes a detailed listing of a large number of the maps from the period 1700 to 1910.
81. Lister (1684).
82. The reference to the map's being the first geologic map is in Ireland (1943), p. 1231, and Beringer (1954), p. 33. For a description of the map, see Campbell (1949).
83. Jarcho (1978).

84. Jarcho (1978), p. 47.

85. Geike (1905), p. 115: "The gifted Frenchman... is thus the father of all the national Geological Surveys." See also Ireland (1943), p. 1233.

86. A full account of Guettard's life and geological contributions is given in Geike (1905), pp. 104–39.

87. Guettard (1746).

88. Reproduced in Kish (1976), pp. 134–36.

89. Ireland (1943), p. 1240.

90. Hensel (1741).

91. J. Klaproth (1823).

92. Eckert (1925), pp. 523–24.

93. The story of Senefelder's accidental discovery is often abridged to the point of being erroneous. In 1796 he wrote out a bill in ink on smooth limestone on which he had been practicing writing for engraving—that is, backward—in the hope that acid would bite into the stone, leaving the writing raised for relief printing such as letterpress. In 1798, while still attempting to perfect that technique, he discovered the principle of lithography while trying to work out a way to transfer right-reading material on paper to the stone as a guide for the raised relief process. See Senefelder (1819), especially pp. 30–32.

94. As a matter of fact, one of the first uses of transfer lithography (writing right-reading on paper and transferring the image to stone) was "At Munich... Paris... St. Petersburgh... all resolutions... etc., agreed to in cabinet meetings are written down on paper by the secretary with chemical ink; in the space of an hour, fifty impressions may be made and distributed at pleasure." Senefelder (1819), p. 256.

95. Hellmann (1895).

96. The first report appeared in the *Journal des Mines* 23 (1808):421. The map was published with Cuvier and Brongniart (1811). An enlarged edition appeared in 1822 and a third edition in 1835. See also Geike (1905), pp. 363–77.

97. Eckert (1925), pp. 248–53, reviews the general history of geological mapping. A good coverage of early geological mapping in Britain, and all of Smith's mapping, is contained in Sheppard (1920). See also Geike (1905), pp. 380–96.

98. Körber (1959); Robinson and Wallis (1967).

99. Merz (1896), p. 206.

100. Horn (1959).

101. Ritter (1806).

102. Schouw (1823*a,b*).

103. W. George (1969).

104. Dupin (1827).

105. Balbi and Guerry (1829).

106. Quetelet (1831).

107. Scrope (1833); Andrews (1966); Kant (1970); Jarcho (1973).

108. Jusatz (1939); Gilbert (1958); Jarcho (1970*b*).

109. Robinson (1955, 1967).

110. Guettard and Monnet (1780).

111. National Society... (1852).

112. Block (1862).

113. G. Engelmann (1964) describes in considerable detail the saga of the

Humboldt–Berghaus–Cotta–Perthes–A. K. Johnston relationships from the time of Humboldt's lectures in physical geography in 1827–28 at the University of Berlin from which his monumental *Kosmos* (Humboldt, 1845–62) ultimately stemmed.

114. An *Atlas zu Alex. v. Humboldt's Kosmos* by T. Bromme containing seventeen thematic maps was published in 1851 in Stuttgart. Engelmann calls it "plagiarism of the worst kind." (G. Engelmann 1970, p. 21.)

115. G. Engelmann (1964), p. 146.

116. G. Engelmann (1970), pp. 17–21.

117. The writings dealing with the life and career of Humboldt are staggering. No attempt has been made to include even a selection of these in the bibliography. A good brief treatment in English is Sinnhuber (1959a).

118. Hall (1976) explores in considerable depth the development of the earth sciences within the context of the general growth of science, particularly in three periods which he called (1) the scientific revolution, 1450–1700; (2) the industrial revolution, 1760–1830; and (3) the mid-nineteenth century, 1830–70.

119. Robinson and Petchenik (1976), pp. 116–23.

120. Halley (1686). See Hellmann (1897), pp. 1–13, and Chapman (1941), p. 3.

121. See Thrower (1969a), pp. 9–15, Thrower (1969b), and Thrower (1978), pp. 202–7 for a detailed description of the map.

122. Thrower (1969a), p. 15.

123. Hellmann (1897), pp. 10–11 (n. 4).

124. Dampier (1699). See Thrower (1969a), pp. 14–15, and Thrower (1969b), pp. 659–61, which reproduce Dampier's maps.

125. Eckert (1925), pp. 110–13.

126. G. Engelmann (1966) includes a reproduction of one of Ritter's sketches on this subject.

127. The *mémoire* is Humboldt (1817a), and the map appeared in Humboldt (1817b). See Körber (1959) and Robinson and Wallis (1967).

128. The translation was by David Brewster, *Edinburgh Philosophical Journal* 3, no. 5 (1820):1–20, and 3, no. 6:256–74; 4, no. 7 (1821):23–37, and 4, no. 8:262–81. For the school atlas see Woodbridge (1826).

129. Eckert (1925), p. 339.

130. The display of each new subject by isolines has regularly been accompanied by the coining of an appropriate new term, and the total number has become overwhelming; see Gully and Sinnhuber (1961). Eckert (1925), p. 357, calls this practice an "isodisease."

131. Eckert (1925), p. 341.

132. Horn (1959).

133. Kämtz's *Lehrbuch* (1831–36) was translated into English: Kaemtz (1845).

134. Hellmann (1897), p. 9, cites Renou.

135. Olsen (1839); Schouw (1839).

136. Petermann (1850).

137. Petermann (1850), p. 123.

138. National Society . . . (1852).

139. Extensive references to developments in meteorology and climatology are found in Eckert (1925) and, because of the common use of isolines in these fields, also in Horn (1959) and Gully and Sinnhuber (1961).

140. For example, Berghaus's *Physikalischer Atlas,* first Abteilung, Meteorology, map 13, dated 1842, shows the distribution of the frequency of thunderstorms by means of isolines with shading in two maps: "Europe in General" and "Bohemia in Particular."

141. Loomis (1846). See Hellmann (1897), pp. 6–7, who indicates that similar maps had been drawn by Brandes in 1816–20 but were not published.

142. Hellman (1897), pp. 7–9.

143. Kircher (1665).

144. Eckert (1925), p. 100.

145. De Vorsey (1976); Eckert (1925), p. 100, suggests that the mapping added nothing to the cartographic development.

146. Franklin (1786).

147. Johnston (1852).

148. Petermann (1850), map 6.

149. Hall (1976), pp. 99–100.

150. The interest and literature on this subject is surprisingly extensive. The best brief summary is in Wallis (1973), pp. 254–58. Another useful source, including reproductions of seven maps, is Hellmann (1895).

151. Taylor (1956), pp. 172–91.

152. Kircher (1654). In the first edition of his *Magnes sive . . .* (1641), he gave directions for making a map of declination. See Wallis (1973), p. 256.

153. This map is reproduced in Hellmann (1895) and Thrower (1969a, b).

154. Curve (-line), *Kurve, courbe,* and *curvas* (adaptations of the Latin *curvus*) have been so used since the latter half of the seventeenth century. Cajori (1919), pp. 216 ff., refers to Leibniz's proposed problem in 1687 of the "isochronous curve," the curve along which a body falls with uniform velocity, and pointed out that all the mathematicians were in correspondence with one another at the time. See also *The Oxford English Dictionary.*

155. Halley obviously knew of Kircher's work on magnetism (Kircher 1654), in which he described such lines, and lines of equal depth on a map were used as early as 1584 (Robinson 1971).

156. The term "isogone" was introduced by C. Hansteen in the early nineteenth century (Hellmann 1895, p. 20). In a comment, dated 1836, in his *Physikalischer Atlas,* p. 12, Berghaus quoted from Hansteen (1819), p. 11, as follows (trans. by author): "Since Kepler's and Newton's times the mathematicians of Europe have all turned their eyes toward the heavens . . . . The earth communicates the movements in its interior by means of the mute language of the magnetic needle and it would . . . be instructive for us to interpret correctly the luminous writings of the aurora. The connection of meteorology with the aurora and accordingly with the magnetic forces stands out clearly; likewise remarkable is the similarity of Humboldt's isotherms and lines of magnetic inclination."

157. Hellmann (1895). See also Eckert (1925), pp. 319–20 for a review of other studies, which by the 1840s had become quite sophisticated.

158. Buache's thematic mapping has been reviewed by Kish (1976). The biographical information has been taken from Walckenaer (1830), pp. 368–71.

159. Buache kept geography and cartography in the family. He married Delisle's daughter in 1729. She died within a few years, and he inherited all of Delisle's plates. In 1746 he married the sister-in-law of Pitrou, inspecteur-général

des Ponts et Chaussées, who had been his first teacher.

160. Guettard (1746). On p. 364 Guettard lauds Buache and publicly thanks him for his help.

161. These maps appeared in the *Mémoires* of the Académie dated 1751 and 1752, which were published in 1755 and 1756. They are reproduced in Kish (1976).

162. Buache (1752).

163. See Kish (1976), where Buache's theory about the structure or framework of the earth's mountain systems is described in more detail. Buache was derided for two centuries after his death for the proposal, but his suggestion, which no doubt came to him from imaginative contemplation of maps, was no more fanciful than was Wegener's hypothesis regarding continental drift, which was also derided.

164. Buache (1752), p. 415. On the question of priority, see Dainville (1970) and Robinson (1976*b*).

165. Desmarest (1753).

166. Füchsel (1762). See Geike (1905), pp. 197–200.

167. Gläser (1775), pp. 49–50.

168. Charpentier (1778), p. 22.

169. Geike (1905), pp. 251–52, 377, 453–56. Ireland (1943).

170. Scrope (1825), map bound after p. 270.

171. Scrope wrote a memoir treating the volcanic region of central France, Scrope (1827), which was accompanied by two maps and many views, elaborately colored.

172. Dainville (1970), p. 389.

173. *Contour:* Dainville (1970) argues that the contour is derived from the isobath. *Hachure:* J. G. Lehmann (1799).

174. Cuvier and Brongniart (1835), pp. 597–607, give a table and a lengthy description of the problem of determining elevations in the vicinity of Paris.

175. Wright (1966), p. 143.

176. Imhof (1965), pp. 3–16, gives a short historical review of the representation of the landform.

177. Du Carla (1782); see also Dainville (1970), pp. 394–95.

178. Map attached to Dupain-Triel (1791) and reproduced in Dainville (1970), fig. 6, p. 397.

179. See G. Engelmann (1966).

180. Société de Géographie de Paris (1822). The announcement of the "Premier sujet de prix" reads: "Déterminer la direction des chaînes de montagnes de l'Europe, leurs ramifications et leurs élévations successives dans toute leur étendue."

181. Under the terms of the competition the materials submitted could bear no names and became the property of the society. The map is preserved in the collection of the Société housed in the Département des Cartes et Plans of the Bibliothèque Nationale in Paris.

182. Olsen (1833). The map is dated as 1830 because, as Olsen stated, the addition of data was going on so rapidly that, what with delays and the time it took to have the engraving done, the map was almost out of date by the time it appeared (pp. vi–vii).

183. Olsen (1833), pp. vii–viii. He further observes that hachures, if done

properly, require just as much accurate data as contours, and that he believes his contours with a 1,000 foot interval give a good general orographic sketch of the continent. Furthermore, he said it would have been ideal to have had both a hachure map and a contour map, but he also said that it would be too costly.

184. Biographical data from Erslaw (1962), 2:496–98.

185. Berghaus (1845), introductory remarks to third Abteilung, map 3, pp. 113–14. See also Johnston (1848), p. 1 of Geology, map 5.

186. For a listing (incomplete) of many such publications, see Wolkenhauer (1916–17).

187. Wolter (1972).

188. Schouw (1823a,b).

189. The maps are interesting technically for the period, since the hemisphere maps were used repetitively as base maps. Lettering unique to a distribution was apparently overprinted.

190. Schouw (1833a,b).

191. Eckert (1925), pp. 388–91.

192. Eckert (1925), p. 412.

193. Arnberger (1966), pp. 79–182, in his section on the history of thematic cartography, defines thematic cartography much more broadly and in consequence cites numerous earlier maps relating to man, such as battle maps, road maps, and such.

194. Crome (n.d.) calculated and symbolized density by the amount of geographical area per 1,000 inhabitants.

195. G. Engelmann (1966), p. 110.

196. Jarcho (1973), pp. 840–44, where a section of the map is reproduced.

197. The date "1830" is immediately below the title, and it more likely refers to the year of the census than to the year the map was made. The populations of the départements are given on the margins, and a comparison with enumerations would, no doubt, reveal if they are indeed those of 1830. If they are from a census taken in 1830, then the date of their availability would at least put an earlier limit on when the map was made.

198. Kant (1970) reproduces plate 2 of Frère de Montizon's map with no reference to plate 1. (See Verner 1974 for cartobibliographic designations.) Plate 2 is obviously later: it is better drawn, the dots are quite evenly spaced (less of a good thing for a dot map), and it contains many more notes.

199. A. or A. J. Frère de Montizon is the way the author gave his name, but the printer-lithographer of his works was always listed simply as A. de Montizon. Perhaps they are different.

200. Du Bus (1931) is the earliest reference to the map of which I know. Since first seeing the maps in 1964 in the Bibliothèque Nationale I have devoted considerable effort toward identifying him, or references to his maps, aided by several agencies and individuals in Paris, especially Mlle Monique Carlier. The mystery has not lessened. He did make one other map, "Carte Statistique de la France Electorale (Chambre de 1839)," Paris, not dated.

201. Prussia (1828). No thorough description of it has been published. It is listed (no. 105, pp. 71–72), as XI. HA, Atlas 175 in the catalog of *Preussen im Kartenbild*, Geheimes Staatsarchiv Preussischer Kulturbesitz, Berlin, 1979.

202. Scrope (1833). The map is described by Andrews (1966) and Jarcho (1973).

203. Scrope (1873), pp. 347–48.
204. Petermann's map in Dieterici (1859).
205. D'Angeville (1836).
206. D'Angeville (1836), p. 15.
207. Railway Commissioners (1838b). It has been asserted that the most certain way to bury a contribution is to publish it in a government document.
208. The circumstances in which the atlas came to be prepared and a detailed description of it are given in Robinson (1955).
209. Harness (1838), p. 41.
210. The term "dasymetric" derives from the technique devised by the Russian Semenov-Tian-Shansky and incorporated in the "Dasymetrischeskaya Karta Europeiskoi Rossii," Leningrad, 1922.
211. The likely individuals are: Commissioner Thomas Drummond, the inventor of the limelight (the Drummond light), who also devised proportional indexes for representation for the Reform Bill in 1831; Commissioner Peter Barlow who was professor of mathematics at the Royal Military Academy at Woolwich (where Harness was an instructor); and Captain Thomas Larcom, superintendent of the Ordnance Survey Office at Phoenix Park in Dublin, where the map was drawn. (The engraving was done in London.) The most probable is Drummond, chairman of the Commission and under secretary of state for Ireland. John Andrews, of Trinity College, Dublin, has called to my attention that in a note on Drummond dated 5 June 1840 by Peter Barlow, to be supplied to Drummond's obituarist (Drummond died in 1840), Barlow wrote: "The three maps illustrative of these chapters (viz. 2 and 3) were formed at the suggestion of Mr. Drummond by Lt. Harness."
212. Ireland (1843), 24:55, plate 1.
213. Ireland (1843), p. xiv. They went on to point out that including towns up to two thousand in the rural population is a source of error in the shading, especially when there was more than one in a district, and observed: "These anomalies, however, do not disturb the general effect of the map as a graphic representation, and the shading on such maps ought never to be considered in any other light."
214. Larcom (1879). An unknown author actually ascribed the maps in the "Railway Atlas" to Larcom: "In 1837 and 1838 he executed the maps already mentioned, with a view to commencing on a scientific plan the railways of Ireland." Larcom (1886), p. 470.
215. Fletcher (1849).
216. Weller (1911).
217. G. Engelmann (1977), pp. 76–79.
218. Stanford (1904).
219. S. Clark (1852). The National Society Atlas does not carry any author designation, but all but one of the fourteen maps are listed as "Designed by the Rev. S. Clark, M.A."
220. E. J. H. Clark (1878).
221. *Dictionary of National Biography*, 4:405–6. Also E. J. H. Clark (1878), p. 234, quotes from a letter written in 1848 by S. Clark: "I have succeeded in some experiments I was having made in coloured lithography for my large maps beyond my hopes."
222. *Dictionary of National Biography*, 4:405–6.

223. Great Britain (1852), vol. 1, facing p. xix.

224. Great Britain (1852), vol. 1, facing p. xlvi.

225. Robinson (1971).

226. Denmark (1857), maps between pp. xvi and xvii in Einleitung, pp. i–xliii. They were reprinted in Denmark (1874).

227. A geographical mile is defined as 1/60 degree of the earth's circumference, a nautical mile.

228. "Marine-Lieutenant Ravn: Bevölkerungskarten über die Dänische Monarchie. 2 bl. im Mst. von 1:1,930,000. Kopenhagen." *Petermanns Mitteilungen* 4 (1858):433, 435.

229. Sydow (1859), p. 255.

230. For example, Petermann (1855) and Petermann (1856).

231. Block (1862). The maps were printed at the Institut Lithog. de Ch. Hellfarth at Gotha, which did all the lithography for *Petermanns Mitteilungen*.

232. Arnberger (1966), pp. 94–97, lists a number of such maps for Austria and central Europe.

233. Klaproth (1823).

234. Eckert (1925), p. 482.

235. Hume (1860).

236. See "Americae Nova Tabula" in Blaeu's *Atlas Novus*, Amsterdam, 1635, vol. 1. Reproduced as plate 56 in R. A. Skelton, *Decorative Printed Maps of the 15th to 18th Centuries* (London, 1952; Spring Books ed., 1965).

237. Eckert (1925), p. 438; C. F. Weiland, "Weltkarte zur Übersicht der vorzüglichsten Varietäten des Menschen" (Weimar, 1835).

238. J. C. Pritchard, "Six Ethnographical Maps with a Sheet of Letterpress" (London, 1843). Eckert (1925), p. 439, points out that Pritchard recognized thirty-two races in Asia, twenty-nine in Europe, thirty-one in Africa, twenty-three in North America, and ten in South America.

239. G. Engelmann (1964), p. 139.

240. A. K. Johnston (1843), p. 53.

241. In the introductory remarks to chart 9, "Ethnographic Map of Great Britain and Ireland," in *The Physical Atlas*, Kombst wrote that his "Ethnogeographic Map of Europe" had been first published in 1841.

242. A. K. Johnston (1843), p. 56, and also in A. K. Johnston (1848). Remarks accompanying Phytology and Zoology, chart 8, p. 4.

243. Postscript to map 9 in A. K. Johnston (1848).

244. A. K. Johnston (1843), pp. 52–53.

245. G. Engelmann (1965).

246. Fick (1971), pp. 132–35.

247. Eckert (1925), pp. 523–24.

248. Arnberger (1966), pp. 135–36. Eckert (1925), p. 524, seems to have made an error in dating the works of Blum.

249. There are occasional intriguing references to unusual maps. For example, Eckert (1925), p. 526, cites a *Nouvelle almanach des gourmands...*, by A. B. de Perigord (Paris, 1825), "with 1 gastronomic map of France." It is likely that a variety of economic maps of smaller areas or special subjects are buried in books and various official reports.

250. Arnberger (1966), p. 142.

251. G. Engelmann (1965), p. 238.
252. G. Engelmann (1965), p. 238.
253. G. Engelmann (1965), p. 239.
254. S. Clark (1852).
255. The map sheet is undated, but it is likely that it is 1851, since the accession stamp of the British Museum is 29 January 1852.
256. Each tiny unique sign was placed on the end of a spokelike shaft radiating from a circle. Petermann's idea apparently was that the shafts, like the spokes of a wheel, could radiate from a central hub and thus characterize the manufactures and materials of a locality. A variety of ways to assemble or combine point symbols referring to the diverse characteristics of places have regularly been devised since Petermann's attempt, none of which seems to have met with much success.
257. Robinson (1967).
258. Robinson (1967), pp. 106-8.
259. Minard (1858) noted that the data were derived from "the excellent statistics of the *consommations de Paris* of Husson, p. 132."
260. Block (1862).
261. Robinson (1955); Railway Commissioners (1838*a*).
262. Harness (1838), p. 442. They are not directly proportional, which may well be the fault of the engraver (Gardner, London). Nothing is known of Harness's original drawing.
263. Rawson (1838).
264. Seyn (1935), p. 44.
265. *Map 1:* "Carte du mouvement des transports en Belgique dressées sur les Données... du Dep$^t$ des Travaux Publics, Carte 1, Année 1834," Ph. Vandermaelen, Bruxelles, no date. *Map 2:* Idem, "Carte 2, Année 1844," same publisher, no date. Each map ca. 987 × 694 mm. Lithograph, hand colored. The maps are not reproduced here, even in part, because of their size.
266. Belpaire (1847), p. 1.
267. "Carte du mouvement des transports en Belgique pendant l'année 1843. Indiquant l'importance comparative de la circulation sur les voies de communications par terre." No publisher, no date, or scale. Map ca. 810 × 628 mm. Lithograph, hand colored.
268. Minard knew of Belpaire's maps, since one of Minard's flow maps (1847) combined the data on Belpaire's 1834 and 1844 maps with a specific reference to them. In his old age Minard wrote that he was pleased "at having given birth in my old age to a useful idea" (Minard 1862, p. 6). Minard's first flow map is dated 1845, when he was aged sixty-four. The statement above was made when he was aged eighty.
269. These are large maps, ca. 650 × 850 mm, with distinctions between railway and waterway traffic made in light green and pink.
270. Minard (1862), p. 3.
271. Robinson (1967), p. 96: Block (1878); Cheysson (1878); Marey (1878).
272. Jarcho (1970*b*).
273. Dupin (1827), plate 1 in vol. 2. See also Du Bus (1931), p. 1. The French *carte figurative* has no precise English equivalent. It generally connotes a noticeable degree of planimetric approximation, and today such a map would

probably be called a *carte thématique*. "Sketch map" suggests a less finished product.

274. On returning to France, Dupin published the result of his survey (1824), which was immediately translated into English (Dupin 1825). The accompanying atlas is entirely concerned with docks, bridges, and such.

275. A short biography of Pierre Charles François Dupin by Paul Tannery appears in *La Grande Encyclopédie*..., Paris, 15:81. He was known as Baron Charles Dupin.

276. Dupin (1827), p. 249, quotes from his address, in which he referred to the map. Funkhouser (1938), p. 300, states that Dupin "published in 1819 a map showing by gradual shadings from black to white the distribution and intensity of illiteracy in France" but gives no evidence for that date or for any other map. I believe he is in error and is referring to the map published in 1827. Both Funkhouser and Beniger and Robyn (1978), p. 10, give that probably apocryphal map a title "Carte de la France Eclairée et de la France Obscure," derived no doubt from Dupin's biography in *La Grande Encyclopédie*... (which gives no date for the map). Those terms for dividing France on the basis of education were apparently introduced after 1819. See note 361.

277. The similarity in design probably stems from the fact that Somerhausen's map was prepared in the lithographic establishment of Jobard Frères, Bruxelles and Amsterdam. Jobard Frères had produced a lithographed version of Dupin's map, which had first been produced by engraving in Paris. The date of Somerhausen's map is probably post-1829. In a note at the bottom is a reference to Quetelet's "...Recherche sur la population..., etc. Bruxelles, page 65 et suivantes." This probably refers to Quetelet (1829).

278. Guerry (1864), note on p. LVI.

279. Hallam (1836). Dupin concluded that births and deaths showed no relation to fluctuations in the price of grain, although the number of marriages seemed to be affected. With respect to ratios of crime, they tended to increase proportionally with density, especially offenses against property. Nothing is known of the map.

280. The date of 1829 does not appear on the sheet but is given by Balbi elsewhere.

281. Biographical details by Maury, pp. 3–7, and Diard, pp. 8–22, in Guerry (1867).

282. Guerry was graphically innovative. For example, he published diagrams and graphs relating meterological variations to physiologic data in Guerry (1829).

283. Biographies of Quetelet are readily available. A useful source is Hankins (1908). See also Merz (1903), pp. 579 ff., and Quetelet (1850).

284. Quetelet (1829).

285. Quetelet (1831, 1832).

286. Quetelet (1831), p. 31.

287. Quetelet (1831), p. 83.

288. Note that Quetelet's portrayal is not a "shaded relief" map in which a three-dimensional surface is shown as though illuminated from a light source.

289. Guerry (1833), pp. III, 24.

290. Greg (1835), p. 63 (italics in original); Moconochie as reported in *Pro-*

*ceedings*, Statistical Society of London, 1, no. 1 (1834–35):8* *(sic)*; Porter (1837).

291. Quetelet (1835). The comment is paraphrased from the publisher's notice of the English translation—Quetelet (1842), p. iii.

292. Quetelet (1838).

293. Quetelet (1842).

294. D'Angeville (1836). See also Du Bus (1931), p. 2.

295. Ireland (1843), 24:xxxiii.

296. Fletcher (1847), p. 195.

297. Fletcher (1849).

298. Fletcher (1849), p. 166. Prince Albert was much interested in statistics (see Quetelet 1850 and Schoen 1938), and he attended the meeting of the Statistical Society at which Fletcher gave his long paper.

299. Mayhew (1862), pp. 449 ff.

300. The complete list is: (1) density of population in London (all other maps are of England and Wales), (2) density of population, (3) intensity of criminality, (4) intensity of ignorance, (5) illegitimate children, (6) early marriages among males, (7) number of females to every 100 males, (8) number committed for rape, (9) number committed for carnally abusing girls, (10) number committed for keeping disorderly houses, (11) number attempting to procure miscarriages, (12) number committed for assaults with intent to ravish and carnally abuse, (13) number committed for bigamy, (14) number committed for abduction, and (15) the criminality of females.

301. Guerry (1864).

302. Guerry (1867).

303. John (1884), pp. 367–68, in contrasting Quetelet and Guerry states that whereas Guerry gathered data in order to make a judgment of the moral character of a region, Quetelet exceeded that and wanted to know the respective levels of populations, and such. Quetelet's assertion that "society contains within itself the seeds of crime" is certainly likely to be more intriguing—to most people—than a shaded map.

304. The analysis of the mechanisms of development, survival, and migrations of pathogens is a fascinating study involving most of the elements found in novels of international intrigue. The background of the several cholera epidemics in western Europe is well told in Siegfried (1960).

Fortunately, and quite in contrast with most other areas of thematic cartography, several competent and dedicated scholars have devoted much effort to the history of medical mapping. Our understanding has been greatly advanced by the researches of a surprising number of students who find this a fascinating topic, but especially by Jarcho, Jusatz, and Gilbert.

305. Jarcho (1970*b*), pp. 131–32.

306. Stevenson (1965).

307. Identification of the cholera bacillus did not occur until 1883, when it was isolated by the German bacteriologist Robert Koch.

308. Jarcho (1970*b*). I know of two others; one of Magdeburg (no author) 1831–32 is in the British Library, and Du Bus (1931), p. 48, cites an 1832 cholera map of Rouen in Hellis, *Souvenirs de cholera en 1832*, Paris, 1833.

309. Rothenburg (1836). Cited also by Jarcho (1970*b*) as in *Zeit. Ges. Med.*, 2 (1836):401–9.

310. Rothenburg (1836), p. 28.

311. Grainger (1850), p. 199.

312. Baker (1833), p. 9.

313. Baker's map given as figure 2 in Gilbert (1958) is not a reproduction of the original but a (neater) redrawing. The original is reproduced in D. Ward, "Living in Victorian Towns," *Geographical Magazine,* May 1971, p. 580, but, inexplicably, green has been overprinted instead of red.

314. The report was given 3 January 1833. On 21 January the Board of Health resolved that the report should be printed, and the title of the map included with the printed report, "Cholera Plan of Leeds Illustrated in this Pamphlet," suggests that it was newly prepared for the printed report.

315. Barry (1837), p. 84.

316. Fortunately many have been saved. Some accompanied elaborate investigations such as Acland's study of Oxford, see Gilbert (1958), pp. 179–82.

317. Perry (1844).

318. Dr. Perry stated that the cause of an epidemic disease is not known. He observed that the general cause is often ascribed to malaria, that is, bad air, (see Jarcho 1970a), or miasma arising from decaying animal and vegetable matter, but he was astute and observed that it did not seem to make any difference when everything was frozen in Glasgow.

319. Quoted from a printed slip bound with Perry (1844).

320. C. F. Weiland, "Heilquellen Karte oder die Eisen, Schwefel, Alkalein, Bittersalz, Glaubersalz, Kochsalz, oder Kohlensäure haltende Mineralwasser, Gaz und Schlammbäder...," Weimar, 1835. Weiland also made a cholera map in 1832 (Eckert 1925, p. 491), and he made a wine map in 1834 (Eckert 1925, p. 525). W. Engelmann (1858), p. 269, gives the date of the wine map as 1843.

321. Jarcho (1974).

322. Michaelis (1843). The map impressed Humboldt sufficiently that he presented it to the Académie in 1845. ("'Correspondance' Séance du lundi, 17 Feb. 1845,'" *Comptes Rendus* 20 [January–June]:450–51.)

323. Hubertz (1854). No map accompanied the printed paper, but a footnote states that it was published in the *Annales Médico-psychologique*.

324. Snow (1855). The first edition was only a slender pamphlet published in 1849. He was awarded a prize of £1,200 by the Institute of France. The second edition was translated into German in 1856.

325. See, for example, Cooper (1853), p. 351, who had done exactly that on a map of Hull he displayed in 1853.

The use of point symbols (dots, crosses, etc.) on maps to show occurrences of disease goes back to the late eighteenth century in the United States, more than fifty years earlier than Snow. Such relatively crude "spot" maps of deaths from yellow fever were utilized in the contagion–anticontagion debate concerning its cause. See Stevenson (1965).

326. Snow (1855). Snow was most ingenious, and he treated his investigation strictly as an experiment where some 300,000 persons were divided into two groups without their choice. Most residents, being tenants, did not know which company supplied their water. Consequently Snow obtained a great many samples and prepared the map by determining their salinities, water from the Thames being relatively more saline.

327. G. Engelmann (1964), p. 145.

328. Jarcho (1969), p. 414, points out that Berghaus's maps are genuine geographical studies verging on the ecological because they included isotherms and other significant environmental data.

329. Jusatz (1940) terms it a model for geomedical maps of epidemics, but he is mistaken in saying that Petermann probably invented the symbolism. As we have seen, Quetelet seems to have been its inventor. He also calls it a *Punktmethode* (dot mapping), but there are no dots with a unit value; it is strictly shading, as Petermann calls it in the "Explanation."

330. Petermann (1848). Petermann's introduction is quoted in full in Gilbert (1958), p. 178.

331. Gilbert (1958) redrew both the cholera map of the British Isles and that of London (Gilbert, figures 4 and 6) to accompany his paper. Figure 4 does not do justice to the original, and figure 6, the map of London, is definitely misleading, since it contains a legend with six black-and-white patterns to symbolize classes. Petermann's original map has no legend, stated classes, or patterns.

332. Ardant (1972), 2:172.

333. Ireland (1843), 24:xiv, xvi–xvii.

334. Larcom (1843).

335. Heywood (1843), p. 282.

336. Wilde (1843).

337. *Athenaeum* (1844).

338. Chadwick (1842).

339. Booth (1889).

340. For example, Block (1861) includes twelve maps of France (by Petermann) on the following subjects: population density; increase and decrease of urban populations since 1856; births in general; illegitimate births; marriages; mortality; creeds; instruction; criminality; lawsuits; wealth; and occupations.

341. Reports from committees (1849). Map bound before p. 9, "Map Exhibiting the Relative Provision of Books in Libraries Publicly Accessible in the Principal States of Europe, as Compared with Their Respective Populations," ca. 390 × 300 mm.

342. Dainville (1964), figure 23 (p. 191) (forests) and plates 22 and 23 (mountains and glaciers).

343. Robinson (1975).

344. Woodward (1975b).

345. In the woodcut technique the nonprinting areas were carved away with a knife on a smooth block whose flat surface paralleled the grain. The display, left in relief, was inked and printed the same way as letterpress. Wood engraving is conceptually the same but was done on the polished end grain of a block using engraver's tools to cut away the nonprinting areas. Wood engravers became incredibly skillful. See Woodward (1975b).

346. Verner (1975). See also Woodward (1978), pp. 163–66.

347. Ristow (1975), pp. 77–104; Senefelder (1819).

348. For a detailed description and analysis of the processes with a useful glossary, see Pearson (1978). See also Koeman (1975).

349. A copperplate engraving is likely to appear a bit sharper, the ink being thicker and adhering to a paper surface that has been somewhat smoothed by

the great pressure required in the copperplate printing process.

350. A case in point is provided by the first choropleth map, Dupin's "Carte figurative de l'instruction populaire de la France." That map was prepared as a copperplate engraving by Adam in the Parisian publication of Dupin's book and as a lithographic engraving (FIG. 75) by Jobard Frères (Bruxelles and Amsterdam). Except for a different kind of water lining in the sea, they are virtually indistinguishable.

351. Pearson (1978), pp. 36–41.

352. Before each impression is taken from a copperplate, the grooves must be filled with ink and then the plate wiped clean. If a "groove" is very wide the ink will be removed in the wiping process. Solid colors can be simulated only by covering an area on the plate with a fine-textured array of small depressions or by closely spaced patterns of rulings.

353. Pearson (1978, 1980).

354. In the nineteenth century a remarkable number of innovations in ways of producing and duplicating maps occurred, but many were short-lived. See Harris (1975), and elsewhere in Woodward (1975a).

355. A brief treatment of the various ways flat tones could be obtained on a copperplate is given in Hind (1923), pp. 9–14. The ruling machine was a mechanical device with which an engraver could obtain a uniform series of closely spaced parallel lines. Commercial engravers used the ruling machine for the sky in artistic prints, but it was of course ideal for putting a flat tint across an area, such as a county or *département*, not encumbered with many other marks.

356. Green (1801).

357. For example, in a three-tone portrayal all but the area to be darkest would be covered with the stop. After the first bite the stop on the second area would be removed. The next application of acid to the entire area would bite the first area deeper and give the second area its first bite. The stop on the third area would then be removed and the whole bitten again, creating a plate that would print three tones.

358. Pearson (1978), pp. 246–49.

359. Pearson (1978), pp. 251–65.

360. In mezzotint one works from dark to light as opposed to employing aquatint, rocker, roulette, or mechanical ruling on the plate, in which one works from light to dark. In the latter processes one starts with a smooth copperplate, which would "print" white, and then grooves or roughens it so that it will print darks. In pure mezzotint one first produces a roughened plate that would print uniformly dark. Lights are then obtained by burnishing down the roughness, to a lesser or greater degree, so that the burnished area can print more or less ink. Burnishing can be employed, of course, in any copperplate engraving process to lighten either tones or lines. The plate may be line engraved before or after tonal effects have been prepared, usually before for flat tones and after for continuous tones.

361. Conrad Malte-Brun, born in Denmark in 1775 but exiled in 1800 for his championing of the French Revolution, was a founder (1821) and first secretary of the Société de Géographie de Paris. Quetelet (1831) credited Dupin with the terms *la France obscure* and *la France éclairée*, but in a letter from Guerry which Quetelet added to his 1831 paper, Guerry pointed out that Malte-Brun had employed those terms as early as 1822 in the *Journal des Débats*.

362. A curious quirk in the history of thematic cartography is that the first makers of choropleth maps tried to make no-class maps but could not do so. After the first decade or so, all choropleth maps employed class limits. Cartographers can now obtain graded tones by computer methods, making no-class maps possible, and they are considering whether they should do it. See, for example, Tobler (1973) and Dobson (1973).

363. Minard devised a no-class system for representing population density in a map of Spain by using patterns of parallel lines spacing them so that a one-centimeter space perpendicular to the lines in a district would include as many lines as there were times five persons per square kilometer in that district. For example, in Murcia one centimeter includes seven lines; $7 \times 5 = 35$ persons per square kilometer. Minard (1867), plate 2. See also Robinson (1967).

Note that Minard's suggestion and some of the modern, computer-produced "tones" are really patterns graded according to ratios of black to white. Flat tones are possible in computer-directed electronic displays, and the hard copy from them, but printer outputs with textures coarser than about seventy-five lines per inch (ca. thirty lines per centimeter) are not seen entirely as visual tonal values and if they are coarser than about forty lines per inch (ca. sixteen lines per centimeter) they are not seen as tonal values at all. See Robinson, et al. (1978), pp. 312–19.

364. Kant (1970).
365. Petermann and Behm (1857).
366. Kant (1970), p. 8, and fig. 3.
367. Hargreaves (1961).
368. Dainville (1958); Destombes (1968).
369. Dainville (1964), pp. 219–22; plates 10–21.
370. J. Wyld, "Chart of the World Shewing the Religion, Population and Civilization of Each Country" (London, 1815). See Jarcho (1973).
371. Harness (1838), p. 41.
372. C. J. Minard, "Carte de l'importance des ports maritime de France mesuré par les tonnage des Navires entrés et sortis en 1850" (Paris, 1852).
373. Robinson (1955), pp. 447–48.
374. W. Bone, "England and Wales, in Divisions and Registration Counties," in *Census of Great Britain, 1851*, vol. 1 (London, 1852), facing p. xix; ditto for Scotland, facing p. xx.
375. Funkhouser (1938), pp. 274–78; Beniger and Robyn (1978).
376. *Allgemeine Deutsche Biographie*, vol. 4 (Leipzig, 1876), pp. 606–7.
377. Crome (1785). The "accompanying illuminated chart" is missing from the copy examined but is described in the foreword.
378. Schilder (1975), nos. 1020, 1021.
379. Funkhouser (1938), pp. 280–81.
380. Playfair (1786).
381. Playfair (1801).
382. Donnant (1805).
383. In the nineteenth century there appears to have been an argument about who invented the "graphic method" in statistics. Block (1878), who makes no reference to Crome, credits Playfair as being the originator as opposed to Donnant. He refers to the question (pp. 380–81) as follows (in translation): "Now, says Peuchet" [Peuchet 1805, p. 33] "this principle of which M. Playfair

speaks as [being] an interesting discovery is to represent each state of Europe by a circular figure in such a manner that their relative power is expressed by the ratio of these figures to one another.... The evidence is conclusive. In effect, Peuchet and Donnant, probably because of professional jealousy, were constantly wrangling, and if Donnant had presented as new a method already known, Peuchet would have seized the occasion to ridicule his ignorance."

384. Carte figurative et approximative de l'importance des ports maritimes de l'Empire Français measurée par les tonnages effectifs des navires entrés et sortis en 1857" (Paris, 1859).

385. Minard (1862), p. 5.

386. The earliest noted use is in the Han maps. See Hsu (1978). See also Dainville (1964), p. 136.

387. Wright (1925), p. 253.

388. J. G. Lehmann (1799). The notion of systematically varying darkness to show differences in slope and elevation had been proposed before Lehmann, but the Saxon army officer was the first to suggest doing so by varying the widths of lines to show the varying angles of slope of the surface. The lines follow *fallinean* or the direction water would flow, that is, normal to the contours. See Imhof (1965), pp. 238–44.

389. Eckert (1921), pp. 509–35.

390. Harness (1838). Quoted in Robinson (1955), pp. 442–43.

391. Robinson (1967).

392. Robinson and Wallis (1967), p. 121.

393. There is an abundant literature on the history of isolines. Special mention should be made of Eckert (1921, 1925), Horn (1959), Gully and Sinnhuber (1961), and Robinson (1971), the first three of which have extensive bibliographies.

394. Andreae and Hoff (1947), p. 19.

395. Riel (1925).

396. Robinson (1976b).

397. Dainville (1958).

398. Du Carla (1782).

399. The application of contouring to large-scale mapping came rather rapidly after 1810, especially in France, where it became official practice. See Dainville (1958).

400. Szaflarski (1959). Szaflarski, as well as other students of the history of cartography, correctly points out that the shaded maps by Zeune (1804) and Ritter (1806) (FIG. 39) are not layer-tint maps, since they have no discernible contours. (See Eckert 1921, p. 458, and Steinhauser 1858, p. 71).

401. Steinhauser (1858), p. 62.

402. Szaflarski (1959), p. 75.

403. Sydow (1838–40).

404. Ireland (1845), appendix 44, part 4, p. iv.

405. Hellmann (1895, 1909); Horn (1959), p. 226.

406. Gully and Sinnhuber (1961) list at least 125.

407. Horn (1959), pp. 229–30.

408. Horn (1959), p. 231.

409. Robinson (1971).

410. Kämtz first published his treatise on meteorology (*Lehrbuch der Meteorologie,* 3 vols.) between 1831 and 1836. The French translation appeared in 1843 (Kämtz 1843) and the English translation was made from the French in 1845 (Kaemtz 1845).
411. See also Beniger and Robyn (1978), p. 5.
412. Lalanne (1845), p. 440.
413. Robinson (1971), pp. 52–53.
414. Petermann (1856).
415. Robinson (1971), p. 52.
416. Sydow (1859), p. 255.

# REFERENCES

THE LITERATURE ON THE EARLY HISTORY OF THEMATIC cartography in western Europe per se is relatively sparse. Although most relevant sources have been included, the following list serves primarily to provide the bibliographic data for all the citations in the Notes. A few useful sources have been added, but most of the obvious reference works, such as biographical dictionaries and encyclopedias, have not been listed. Only those thematic maps reproduced herein and which were published as separates have been included.

Adams, F. D. 1938. *The birth and development of the geological sciences.* New York: Williams and Wilkins (reprinted New York: Dover Publications, 1954).

Adler, B. F. 1910. *Maps of primitive peoples* (Saint Petersburg, in Russian). *Bulletin* of the Imperial Society of Students of Natural History, Anthropology and Ethnography, at the Imperial University of Moscow. Vol. 119. Works of the Geographical Section, no. 2. See de Hutorowicz (1911).

Andreae, S. J. F., and van't Hoff, B. 1947. *Geschiedenis der Kartographie van Nederland.* The Hague: Martinius Nijhoff.

Andrews, J. H. 1966. An early world population map. *Geographical Review* 56:447–48.

d'Angeville, A. 1836. *Essai sur la statistique de la population française considérée sous quelque-uns de ses rapports physiques et moraux.* Paris.

Ardant, G. 1971–72. *Histoire de l'impôt,* 2 vols. Paris: Fayard.

Armytage, W. H. G. 1965. *The rise of the technocrats.* London: Routledge and Kegan Paul.

Arnberger, E. 1966. *Handbuch der thematischen Kartographie.* Vienna: Franz Deuticke.

Arnhold, H. 1968. Medizinische Geographie—eine Auswahlbibliographie. In *Wissenschaftliche Veröffentlichungen des Deutschen Instituts fur Länderkunde,* ed. E. Lehmann, New fol. no. 25/26, pp. 351–484. Leipzig.

*Athenaeum, The*. 1844. Reviews: Census of the population of Ireland in 1841. No. 846 (London, 13 January), p. 29.

Bagrow, L. 1943. The origin of Pholemy's [sic] Geographia. *Geografiska Annaler,* vols. 3–4, pp. 319–87.

Bagrow, L., and Skelton, R. A. 1964. *History of cartography.* London: Watts.

Baker, R. 1833. *Report of the Leeds Board of Health.* Leeds.

Balbi, A., and Guerry, A. M. 1829. (Map.) Statistique comparée de l'état de l'instruction et du nombre des crimes dans les divers Arrondissements des Académies et des Cours R.$^{\text{les}}$ de France. Paris: Jules Renouard. Date not on map but given by Balbi elsewhere.

Barry, D. 1837. Statistics of epidemic cholera. *Transactions of the Statistical Society of London,* 1 (Pt. 1):83–96. (Paper read 15 June 1835.)

Belpaire, A. 1847. *Notice sur les cartes du mouvement des transports en Belgique.* Bruxelles: Ph. Vandermaelen.

Beniger, J. R., and Robyn, D. L. 1978. Quantitative graphics in statistics: A brief history. *American Statistician* 32 (no. 1):1–11.

Berghaus, H. 1845, 1848. *Physikalischer Atlas oder Sammlung von Karten. . . .* Gotha: Justus Perthes.

Berthaut, H. 1898. *La carte de France 1750–1898.* 2 vols. Paris: Service Géographique de l'Armée.

———. 1902. *Les ingénieurs géographes militaires 1624–1831.* 2 vols. Paris: Service Géographique de l'Armée.

Beste, G. 1578. *A true discourse of the late voyages . . . under the conduct of Martin Frobisher. . . .* London. Issued by the Hakluyt Society, first series, no. 38 (1867); reprinted in Richard Collinson, *The three voyages of Martin Frobisher.* New York:Burt Franklin, n.d.

Block, M. 1861. *Bevölkerung des Französischen Kaiserreichs in ihren wichtigsten statistischen Verhältnissen dargestellt.* Gotha: Justus Perthes.

———. 1862. *Puissance comparé des divers états de l'Europe—Atlas.* Gotha: Justus Perthes.

———. 1878. *Traite théoretique et practique de statistique.* Paris: Guillaumin.

Booth, C. 1889. *Life and labour.* Vol. 1. *East London.* London and Edinburgh: Williams and Nargate.

———. 1891–1903. *Life and labour of the people in London.* 17 vols. London.

Brown, L. A. 1949. *The story of maps.* Boston: Little, Brown.

Bruinsz, P. 1584. (Map.) Bescrivinge der diepte vant sparen. . . . Manuscript, n.p.

Buache, P. 1752. Essai de géographie physique. . . . Dated 15 November 1752. *Mémoires de l'Académie des Sciences,* (Année 1752). Paris 1756, pp. 399–416.

Büsching, A. F. 1762. *A new system of geography. . . .* 6 vols. Translated from the German original by A. Millar, maps by Thos. Kitchin. London.

Cajori, F. 1919. *A history of mathematics.* New York: Macmillan.

Campbell, E. M. J. 1949. An English philosophico-chorographical chart. *Imago Mundi* 6:79–84.

Cannon, S. F. 1978. *Science in culture: The early Victorian period.* Folkestone: Dawson; New York: Science History Publications.

Castner, H. W. 1980. Special purpose mapping in eighteenth century Russia: A search for the beginnings of thematic mapping. *American Cartographer* 7:163–75.

Chadwick, E. 1842. *Report on the sanitary condition of the labouring population of Great Britain to the poor law commissioners, 1842.* (HL-), vol. 26. Maps follow p. 160.

Chapman, S. 1941. Edmond Halley as physical geographer and the story of his charts. *Occasional Notes,* no. 9 (Royal Astronomical Society, London).

Charpentier, J. F. W. 1778. *Mineralogische Geographie der Chursächsischen Lande.* Leipzig.

Cheysson, E. 1878. *Les méthodes de statistique graphique à l'Exposition Universelle de 1878. Rapport à la Commission Permanente du Congrès International de Statistique.* Paris.

Clark, E. J. H. 1878. *Memorials from journals and letters of Samuel Clark, M.A., F.R.G.S.* London: Macmillan.

Clark, S. 1852. *Maps illustrative of the physical, political, and historical geography of the British Empire....* London: National Society for Promoting the Education of the Poor. N.d. but probably 1852.

Cooper, H. 1853. On the cholera mortality in Hull during the epidemic of 1849. *Journal of the Statistical Society of London* 16:347–51.

Crome, A. F. W. (N.d.). Verhaeltniss-Karte von den Deutschen Bundesstaaten, zur Übersicht und Vergleichung des Flächenraums, der Bevölkerung.... Friedr. Wilh. III zugeeignet von dem Verfasser.... N.p.

———. 1782. (Map.) Neue Carte von Europa welche die merkwürdigsten Producte und vornehmsten Handelsplätze nebst den Flächen-Inhalt aller Europäischen Länder in deutschen Quadrat-Meilen enthält. Dessau.

———. 1785. *Ueber die Grosse und Bevölkerung der sämtlichen europaischen Statten: Ein Beytrag zur Kenntniss der Stattenverhältnisse und zur Erklärung der neuen Grözen-Karte von Europa.* Leipzig: Weygandschen Buchhandlung.

Crone, G. R. 1966. *Maps and their makers.* 3d ed. New York: Capricorn Books.

———. 1978. *Maps and their makers.* 5th ed. Folkestone and Hamden, Conn.: Dawson–Archon Books.

Crosland, M. 1967. *The society of Arcueil: A view of French science at the time of Napoleon I.* Cambridge: Harvard University Press.

Cuvier, G., and Brongniart, A. 1811. *Essai sur la géographie minéralogique des environs de Paris, avec une carte géognostique, et des coupes de terrain.* Paris: Baudoin.

———. 1835. *Description géologique des environs de Paris.* 3d ed. Accompanying separate atlas. Paris.

Dainville, F. de. 1958. De la profondeur a l'altitude. In *Le navire et l'économie maritime du moyen âge au EXIII<sup>e</sup> siècle principalement en Méditerranée*, pp. 195–213. Paris: Bibliothèque générale de l'Ecole Pratique des Hautes Etudes, VI<sup>e</sup> section. Also published as, From the depths to the heights (trans. Arthur H. Robinson), *Surveying and Mapping* 30 (1970):389–403.

———. 1964. *Le langage des géographes*. Paris: Editions A. et J. Picard.

———. 1968. Carte des places protestantes en 1620, dessinées à la fin règne de Louis XIII. *Journal des Savants*, October–December, pp. 214–43.

Daly, C. P. 1879. On the early history of cartography; or, What we know of maps and map-making, before the time of Mercator. *Bulletin of the American Geographical Society* 11:1–41.

Dampier, W. 1699. Discourse of trade-winds, breezes, storms, seasons of the year, tides and currents of the torrid zone throughout the world. In *Voyages and descriptions*, 2 (part 3):1–112. London.

De Hutorowicz, H. 1911. Maps of primitive peoples. *Bulletin of the American Geographical Society* 43:669–79. A brief abridgment of Adler (1910).

Denmark. 1857. *Statistischen Tabellenwerk*. N.s., vol. 12 (German edition). Copenhagen: Statistical Bureau.

———. 1874. *Tillag til Statistik Tabelwaerke*. Tredie Raekke, Attende Bind, indeholdende Populationskaart over Kongeriget Danmark for Aaret 1870 samt for Aarene 1855 og 1845. Copenhagen: Statistical Bureau.

Desmarest, N. 1753. *Dissertation sur l'ancienne jonction de l'Angleterre à la France*. Amiens.

Destombes, M. 1968. Les plus anciens sondages portés sur les cartes nautiques aux XVI<sup>e</sup> et XVII<sup>e</sup> siècles. *Bulletin de l'Institut Océanographique*, special no. 2 (Monaco), pp. 199–222.

De Vorsey, L. 1976. Pioneer charting of the Gulf Stream: The contributions of Benjamin Franklin and William Gerard de Brahm. *Imago Mundi* 28 (Vol. 2 of second series):105–20.

Dieterici, C. F. W. 1859. Die Bevölkerung der Erde, nach ihren Totalsummen, Racen-Verschiedenheiten und Glaubensbekentnissen. *Petermanns Mitteilungen* 5:1–19.

Dingle, H. 1972. Physics in the eighteenth century. In Ferguson (1972), pp. 28–46.

Dobson, M. W. 1973. Choropleth maps without class intervals? A Comment. *Geographical Analysis* 3:358–60.

Donnant, D. F. 1805. *Statistical account of the United States of America*. Translated from the French by William Playfair... illustrated by a divided circle, representing the proportional extent of the different states... by a new method.... London.

Du Bus, C. 1931. *Démocartographie de la France, des origines à nos jours*. Paris: Libraire Félix Alcan.

Du Carla. 1782. *Expression des nivellemens, ou méthode nouvelle pour marquer rigoureusement sur les cartes terrestres & marines les hauteurs & les configurations du terrein.* Paris: Dupain-Triel père.

Dupain-Triel, J. I. 1791. *Recherche géographique sur les hauteurs des plaines du Royaume.* Paris: Hérault.

———. 1791. (Map.) La France considérée dans les différentes hauteurs de ses plains, ouvrage spécialement destiné à l'instruction de la jeunesse. Paris.

———. 1798–99. (Map.) Carte de la France où l'on a essayé de donner la configuration de son territoire, par une nouvelle méthode de nivellements. Paris.

Dupin, C. 1825. *The commercial power of Great Britain.* 2 vols. Translated from the French; accompanying atlas. London.

———. 1827. *Forces productives et commerciales de la France.* 2 vols. Paris: Bachelier.

Eckert, M. 1921, 1925. *Die Kartenwissenschaft.* 2 vols. Berlin and Leipzig: Walter de Gruyter.

Eckert-Greifendorff, M. 1938. Die Revolutionen in der Kartographie. *Zeitschrift für Erdkunde* 6:785–98.

Engelmann, G. 1964. Der Physikalischer Atlas des Heinrich Berghaus und Alexander Keith Johnstons Physical Atlas. *Petermanns Mitteilungen* 108:133–49.

———. 1965. Frühe thematische Karten zur ökonomische Geographie. *Mitteilungen der geographischen Gesellschaft der Deutschen Democratischen Republik* 33:233–40.

———. 1966. Carl Ritters "Sechs Karten von Europa," mit einer Abbildung. *Erdkunde* 20 (part 2):104–10.

———. 1969. Zeittafel der Kartographie 1700–1850. In *Geographisches Taschenbuch*, pp. 1–20. Wiesbaden: Franz Steiner Verlag.

———. 1970. Alexander von Humboldts kartographische Leistung. *Wissenschaftliche Veröffentlichungen* des Geogr. Inst. der Deutschen Akademie der Wissenschaft, n.s. 27–28, pp. 5–21.

———. 1977. Heinrich Berghaus der Kartograph von Potsdam. *Acta Historica Leopoldina*, no. 10 (cover date given as 1976; title page as 1977). Halle [Salle]: Deutsche Akademie der Naturforscher Leopoldina.

Engelmann, W. 1858. *Bibliotheca geographica.* Leipzig.

Erslaw, T. H. 1962. *Almindeligt Forfatter-Lexicon for Kongerigt Danmark ... fra før 1814 til efter 1858.* Copenhagen.

Fallati, J. 1843. *Einleitung in der Wissenschaft der Statistik.* Tübingen: H. Laup.

Farr, W. 1852. Influence of elevation on the fatality of cholera. *Journal of the Statistical Society of London* 15:155–83.

Fay, B. 1932. Learned societies in Europe and America in the eighteenth century. *American Historical Review* 37:255–66.

Fenniman, J. D. 1974. The Olsen-Bredsdorff map and the adoption of contours for relief depiction on atlas maps in the early nineteenth century. Master's thesis, University of Wisconsin–Madison.

Ferguson, A., ed. 1972. *Natural philosophy through the eighteenth century and allied topics.* London: Taylor and Francis. (First published in 1948 as a supplement to *The Philosophical Magazine*.)

Fick, K. E. 1971. Die Kartographische Darstellung wirtschaftsgeographische Sachverhalte im 18 Jahrhundert. *Geographische Zeitschrift* 59 (no. 2):131–39.

Finke, L. L. 1792–93. *Versuch einer allgemeinen medizinisch-praktischen Geographie*. 3 vols. Leipzig.

Fleming, W. 1955. *Arts and ideas*. New York: Holt, Rinehart and Winston.

Fletcher, J. 1847. Moral and educational statistics of England and Wales. *Journal of the Statistical Society of London* 10:193–221.

———. 1849. Moral and educational statistics of England and Wales. *Journal of the Statistical Society of London* 12:151–76, 189–335.

Foncin, M. 1961. Dupin Triel [sic] and the first use of contours. *Geographical Journal* 127:553–54.

Franklin, B. 1786. A letter from Dr. Franklin, to Mr. Alphonsus le Roy, member of Several Academies at Paris. Containing sundry maritime observations. *Transactions of the American Philosophical Society* 2:294–329.

Füchsel, G. C. 1762. Historia terrae et maris, ex historia Thuringiae per montium descriptionem erecta. *Transactions of the Electoral Society of Mayence* 2:44–209.

Funkhouser, H. G. 1938. Historical development of the graphical representation of statistical data. *Osiris* 3:269–404.

Geike, A. 1905. *The founders of geology*. 2d ed. London: Macmillan; reprinted New York: Dover, 1962.

George, A. J. 1938. The genesis of the Académie des Sciences. *Annals of Science* 3:373–80.

George, W. 1969. *Animals and maps*. Berkeley and Los Angeles: University of California Press.

Gilbert, E. W. 1958. Pioneer maps of health and disease in England. *Geographical Journal* 124:172–83.

Gläser, F. G. 1775. *Versuch einer Mineralogischen Beschreibung der Gerfürsteten Graffschaft Henneberg Chursächssischen Anteils....* Leipzig.

Grainger. 1850. *Report of the General Board of Health on the epidemic cholera of 1848 and 1849: Appendix B, Report by Mr. Grainger*. Parliamentary Papers. Session 1850, vol. 21.

Great Britain. 1852. *Census of Great Britain, 1851*. London.

Green, J. H. 1801. *The complete aquatinter: Being the whole process of etching and engraving in Aquatinta;...etc.* London.

Greenhood, D. 1944. The first graphic art. *News Letter of the American Institute of Graphic Arts*, no. 78, pp. 1–3.

Greg, W. R. 1835. Abstract of a paper on the social statistics of the Netherlands given at the British Association meeting in Dublin (August 1835). *Proceedings of the Statistical Society of London* 1 (no. 3, 1835–36):63–65.

Guerry, A. [M.] 1829. Tableau des variations météorologiques comparées aux phénomènes physiologiques.... *Annales de Hygiène* 1:228–34.

Guerry, A. M. 1833. *Essai sur la statistique morale de la France, précédé d'un rapport à l'Académie des Sciences, par MM. Lacroix, Silvestre et Girard.* Paris: Chez Crochard.

———. 1864. *Statistique morale de l'Angleterre comparée avec la statistique morale de la France* [with] *Atlas. Cartes et constructions graphiques représentant les résultats généraux des tableaux numériques avec une introduction contenant l'histoire de l'application des nombres aux science morale.* Paris: Baillière.

[Guerry, A. M.]. 1867. *Guerry (André-Michel)* [biography]. Paris: Baillière.

Guettard, J. E. 1746. Mémoire et carte minéralogique sur la nature & la situation des terreins qui traversent la France & l'Angleterre. Dated 19 February 1746. *Mémoires de l'Académie des Sciences,* Année 1746, Paris, 1751, pp. 363–92.

Guettard, J. E., and Monnet. 1780. *Atlas et description minéralogique de la France.* Paris.

Gully, J. L. M., and Sinnhuber, K. A. 1961. Isokartographie: Eine terminologische Studie. *Kartographische Nachrichten* 11 (no. 4):89–99.

Gumprecht, T. E. 1856. Die Städte Bevolkerung von Spanien. *Petermanns Mitteilungen* 2:393–99.

Hall, D. H. 1976. *History of the earth sciences during the scientific and industrial revolutions.* Amsterdam: Elsevier Scientific Publishing Co.

Hallam, H. 1836. An abstract of the proceedings of the Statistical Section of the British Association for the Advancement of Science, at the meeting held at Bristol in the month of August, 1836. *Proceedings of the Statistical Society of London* 1 (no. 8):187–90.

Halley, E. 1686. An historical account of the trade winds, and monsoons, observable in the seas between and near the tropicks; with an attempt to assign the phisical cause of said winds. *Philosophical Transactions,* no. 183, pp. 153–68. The issue was published in 1688.

———. 1701. (Map.) A new and correct chart shewing the variations of the compass in the western and southern oceans.... London.

———. 1702. (Map.) A new and correct sea chart of the whole world shewing the variations of the compass as they were found in the year MDCC. London.

———. 1715. (Map.) A description of the passage of the shadow of the moon over England, in the total eclipse of the SUN, on the 22d day of April 1715 in the morning. London.

Hankins, F. H. 1908. Adolphe Quetelet as statistician. *Studies in History, Economics and Public Law* (Columbia University) 31 (no. 4):446–576.

Hansteen, C. 1819. *Untersuchungen über den Magnetismus der Erde* (with separate atlas). Christiania.

Happel, E. G. 1687. *Mundus Mirabilis Tripartitus oder Wunderbare Welt in einer Kurtzen Cosmographia fürgestellet:* ... Ulm: Matthaeus Wagner.

Hargreaves, R. P. 1961. The first use of the dot technique in cartography. *Professional Geographer* 13 (no. 5):37–39.

Harness, H. D. 1838. Report from Lt. H. D. Harness, R.E., explanatory of the principles on which the population, traffic and conveyance maps have been constructed. In Railway Commissioners (1838a), appendix 3.

Harris, E. M. 1975. Miscellaneous map printing processes in the nineteenth century. In Woodward (1975a), pp. 113–36.

Hellmann, G. 1895. Die ältesten Karten der Isogonen, Isoklinen, Isodynamen. *Neudrucke von Schriften und Karten über Meteorologie und Erdmagnetismus*, no. 4. Berlin: Asher.

———. 1897. Meteorologische Karten. *Neudrucke von Schriften und Karten über Meteorologie und Erdmagnetismus*, no. 8. Berlin: Asher.

———. 1899. The beginnings of magnetic observations. *Terrestrial Magnetism and Atmospheric Electricity* 4:73–86.

———. 1909. Magnetische Kartographie in historisch-kritischer Darstellung. *Veröff. des Königlich Preuss. Met. Instituts* (Berlin), no. 215, vol. 3, no. 3, pp. 5–61.

Hensel, G. 1741. *Synopsis universae Philologiae, in qua ... unitas et harmonia linguarum totius orbis terrarum ....* Nürnberg.

Heywood, J. 1843. Thirteenth Meeting of the British Association... Account of the Proceedings of the Statistical Section. *Journal of the Statistical Society of London* 6:281–82.

Hind, A. M. 1923. *A history of engraving and etching from the fifteenth century to the year 1914.* New York: Houghton Mifflin; reprinted New York: Dover, 1963.

Horn, W. 1959. Die Geschichte der Isarithmenkarten. *Petermanns Mitteilungen* 103:225–32.

Howse, D., and Sanderson, M. 1973. *The sea chart.* New York: McGraw-Hill.

Hsu, M.-L. 1978. The Han maps and early Chinese cartography. *Annals of the Association of American Geographers* 68 (no. 1):45–60.

Hubertz, J. R. 1853. Statistics of medical diseases in Denmark, according to the census of July 1st, 1847. *Journal of the Statistical Society of London* 16:244–74.

Humboldt, A. de. 1817a. Des lignes isothermes et de la distribution de la Chaleur sur le globe. *Mémoires de Physique et Chimie, de la Société d'Arcueil* (Paris) 3:462–602.

———. 1817b. Sur les lignes isothermes. Par A. de Humboldt (Extrait). *Annales de chimie et de physique* 5:102–11. The map is folded between pp. 112 and 113.

Humboldt, A. v. 1845–62. *Kosmos: Entwurf einer physischen Weltbeschreibung.* 5 vols. Stuttgart and Tübingen: Cottasche Buchhandlung.

Hume, A. 1860. *Remarks on the census of religious worship for England and Wales ... and a map illustrating the religious condition of the country.* London: Longman, et al.; Liverpool: Holdon.

Imhof, E. 1965. *Kartographische Geländedarstellung.* Berlin: Walter de Gruyter.

International Geographical Union. 1964. *Monumenta cartographica vetustiores aevi.* Under the direction of R. Almagià and M. Destombes. Amsterdam: N. Israel.

Ireland. 1843. *Report of the commissioners appointed to take the census of Ireland for the year 1841* . . . . (Reports from Commissioners, Session 2 February–24 August 1843.) Dublin: Alexander Thom (for HMSO).

———. 1845. Index to minutes of evidence taken before Her Majesty's commissioners of inquiry into the state of the law and practice in respect to the occupation of land in Ireland. Part 5 of *Occupation of land (Ireland): reports from commissioners 1845*. (9th of 14 vols.) Dublin: Alexander Thom (for HMSO).

Ireland, H. A. 1943. History of the development of geologic maps. *Bulletin of the Geological Society of America* 54 (September):1227–80.

Ivins, W. M. 1953. *Prints and visual communication*. Cambridge: Harvard University Press.

Jarcho, S. 1969. The contributions of Heinrich and Hermann Berghaus to medical cartography. *Journal of the History of Medicine and Allied Sciences* 24:412–15.

———. 1970a. A cartographic and literary study of the word *malaria*. *Journal of the History of Medicine and Allied Sciences* 25:31–39.

———. 1970b. Yellow fever, cholera, and the beginnings of medical cartography. *Journal of the History of Medicine and Allied Sciences* 25:131–42.

———. 1973. Some early demographic maps. *Bulletin of the New York Academy of Medicine* 49:837–44.

———. 1974. An early medicostatistical map (Malgaigne, 1840). *Bulletin of the New York Academy of Medicine* 50:96–99.

———. 1978. Christopher Packe (1686–1749): Physician-cartographer of Kent. *Journal of the History of Medicine and Allied Sciences* 33:47–52.

John, V. 1884. *Geschichte der Statistik*. 2 vols. Stuttgart.

Johnston, A. K. 1843. *The national atlas of historical, commercial and political geography . . . accompanied by maps and illustrations of the physical geography of the globe by Dr. Heinrich Berghaus . . . and an ethnographic map of Europe by Dr. Gustaf Kombst.* . . . Edinburgh.

———. 1848. *The physical atlas* . . . . Edinburgh: W. and A. K. Johnston.

———. 1852. *Atlas of physical geography, illustrating, in a series of original designs, the elementary facts of geology, hydrology, meteorology, and natural history*. Edinburgh and London: Blackwood.

Jusatz, H. J. 1939. Zur Entwicklungsgeschichte der medizinisch-geographischen Karten in Deutschland. *Mitteilungen des Reichsamts für Landesaufnahme* 15 (no. 1):11–22.

———. 1940. Die geographisch-medizinische Erforschung von Epidemien. *Petermanns Mitteilungen* 86:201–4.

———. 1969. Medical mapping as a contribution to human ecology. *Bulletin*, Geography and Map Division, Special Libraries Association, no. 78 (December), pp. 19–23.

Kämtz, L. F. 1831–36. *Lehrbuch der Meteorologie*. 3 vols. Halle.

———. 1843. *Cours complet de météorologie*. Trans. Ch. Martins. Paris.

———. Kaemtz, L. F. 1845. *A complete course of meteorology*. Trans. C. V. Walker.

Kant, E. 1970. Über die ersten absoluten Punktkarten der

Bevölkerungsverteilung. *Lund Studies in Geography,* ser. B, *Human Geography,* no. 36, pp. 1–12.

Kircher, A. 1654. *Magnes sive de arte magnetica opus tripartitum.* Rome. (First edition 1641.)

———. 1665. *Mundus subterraneus in XII libros digestus....* Amsterdam.

Kish, G. 1976. Early thematic mapping: The work of Philippe Buache. *Imago Mundi* 28 (vol. 2, 2d ser.):129–36.

Klaproth, J. 1823. *Asia Polyglotta nebst Sprachatlas.* Paris.

Koeman, C. 1975. The application of photography to map printing and the transition to offset lithography. In Woodward (1975a), pp. 137–55.

Körber, H.-G. 1959. Bemerkungen über die Erstveröffentlichung der schematischen Jahresisothermenkarte Alexander von Humboldt. *Forschungen und Fortschritte* 33:355–58.

Kramer, G. 1875. *Carl Ritter, ein Lebensbild.* 2d ed. Halle.

Kutscheit, J. V. 1845. (Map.) Kirchenkarte von Deutschland. Berlin.

Lalanne, L. 1845. Remarques à l'occasion du mémoire de M. Morlet sur les centres de figures; et réflexions sur la représentation graphique de divers éléments relatif à la population. *Comptes Rendus... de l'Académie des Sciences* 20:438–41.

Larcom, T. A. (Capt.). 1843. Observations on the census of Ireland in 1841. *Journal of the Statistical Society of London* 6:323–51.

———. 1879. [No author]. Obituary of Sir Thomas Larcom, *Proceedings of the Royal Society* 29:10–15.

———. 1886. [No author]. A century of Irish government. *Edinburgh Review or Critical Journal* 164:447–83.

Lehmann, E. 1959. Carl Ritters Kartographische Leistung. *Die Erde* 90 (no. 2):184–222.

Lehmann, J. G. 1799. *Darstellung einer neuen Theorie der Bergzeichnung der schiefen Flächen im Grundriss oder der Situationszeichnung der Berge.* Leipzig.

Lister, M. 1684. An ingenious proposal for a new sort of map of countries, together with tables of sands and clays.... *Philosophical Transactions* 14:739 ff. (pagination mixed).

Loomis, E. 1846. On two storms which were experienced throughout the United States, in the month of February, 1842. *Transactions of the American Philosophical Society* 9:161–84.

McCleary, G. F. 1969. *The dasymetric method in thematic cartography.* Ph.D. diss., University of Wisconsin–Madison.

McKie, D. 1972a. The scientific periodical from 1665 to 1798. In Ferguson (1972), pp. 122–32.

———. 1972b. Scientific societies to the end of the eighteenth century. In Ferguson (1972), pp. 133–43.

Malgaigne, J.-F. 1840. Recherches sur la fréquence des hernies, selon les sexes, les âges, et relativement à la population. *Annales d'Hygiène Publique et de Médecine Légale* 24:1–54.

Marey, E. J. 1878. *La méthode graphique dans les sciences expérimentales et principalement en physiologie et en médecine.* Paris.

Mayhew, H. 1862. *London labour and the London poor: A cyclopaedia of the conditions and earnings of those that* will *work, those that* cannot *work, and those that* will not *work.* Vol. 4. London: Griffen, Bohn.

Meinardus, W. 1899. Die Entwicklung des Karten der Jahres-Isothermen von Alexander von Humboldt bis auf Heinrich Wilhelm Dove. In *Wissenschaftliche Beiträge zum Gedächtnis der hundertjährigen Wiederkehr des Antritts von Alexander von Humboldt's Reise nach Amerika am 5 Juni 1799.* Berlin: Humboldt-Centenar-Schrift.

Merz, J. T. 1896, 1903. *A history of European thought in the nineteenth century.* Vols. 1 and 2. Edinburgh and London: Blackwood.

Michaelis, E. H. 1843. (Map.) Skizze von der Verbreitung des Cretinismus im Canton Aargau. Aarau.

Milham, W. I. 1945. *Time and timekeepers.* New York: Macmillan.

Minard, C. J. 1858. (Map.) Carte figurative et approximative des quantités de Viandes de Boucherie envoyées à Paris. Paris.

———. 1859. (Map.) Carte figurative et approximative de l'importance des portes maritimes de l'empire français mesurée par les tonnages effectifs des navires entrés et sortis en 1857. Paris.

———. 1861. (Map.) Carte approximative de l'étendue des marchés des houilles et coke étranger dans l'empire français en 1858. Paris.

———. 1862. *Des tableaux graphiques et des cartes figuratives.* Paris: Thunot.

———. 1864. (Map.) Carte figurative et approximative des poids des bestiaux venus à Paris sur les Chemins de fer en 1862. Paris.

———. 1865. (Map.) Carte figurative et approximative du mouvement des voyageurs sur les principaux chemins der fer de l'Europe en 1862. Paris.

———. 1865? (Map.) Carte figurative et approximative des quantités de vin français exportés par mer en 1864. Paris. (The map shows no date.)

———. 1867. *Appendice à la carte des voyageurs sur les chemins de fer d'Europe en 1862 de M. Minard suivi de considerations sur les chemins de fer.* Paris: Thunot.

Montizon, F. de. 1830. (Map.) Carte philosophique figurant la population de la France. Paris.

Mood, F. 1946. The rise of official statistical cartography in Austria, Prussia, and the United States, 1855–1872. *Agricultural History* 20:209–25.

Myres, J. L. 1896. An attempt to reconstruct the maps used by Herodotus. *Geographical Journal* 8:605–31.

National Society for the Education of the Poor in England and Wales. 1852? *Maps illustrative of the physical, political and historical geography of the British Empire.* London.

Olsen, O. N. 1833. *Commentaire à l'esquisse orographique de l'Europe.* Copenhagen: Imprimerie de Bianco Luno et Schneider.

———. 1839. *Atlas pour le tableau du climat d'Italie.* Copenhagen: Chez Gyldendal.

Olsen, O. N., and Bredsdorff, J. H. 1824. (Map.) Carte orographique de l'Europe pour servir à la connaissance des principales classes de montagnes de leur connexion, et de leur ramification, et de leur hateur relative. (No author, date, or place shown.)

———. 1833. (Map.) *Esquisse orographique de l'Europe par J. H. Bredsdorff & O. N. Olsen, en 1824; corrigée et considerablement augmentée par O. N. Olsen, en 1830, gravée par P. J. Seehusen.* Copenhagen: Reitzel.

Pearson, K. S. 1978. *Lithographic maps in nineteenth century geographical journals.* Ph.D. diss., University of Wisconsin–Madison.

———. 1980. The nineteenth-century color revolution: Maps in geographical journals. *Imago Mundi* 32:9–20.

Perry, R. 1844. *Facts and observations on the sanitary state of Glasgow during the last year; with statistical tables of the late epidemic, shewing the connection existing between poverty, disease, and crime.* Glasgow.

Petchenik, B. B. 1979. From place to space: The psychological achievement of thematic mapping. *American Cartographer* 6:5–12.

Petermann, A. 1848. (Map.) Cholera map of the British Isles, showing the districts attacked in 1831, 1832, and 1833. London.

———. [1848]. *Statistical notes to the cholera map of the British Isles showing the districts attacked in 1831, 1832, and 1833. . . .* London: John Betts.

———. [1849]. (Map.) British Isles, elucidating the distribution of the population based on the census of 1841. . . . London.

———. 1850. *Atlas of physical geography with descriptive letterpress by Thomas Milner.* London: Wm. S. Orr.

———. 1855. Graphische Darstellungen der Bevölkerungs-Verhältnisse . . . in den Vereinigten Staaten von Nord Amerika, nach dem neuesten Census (1850). *Petermanns Mitteilungen* 1:128–42; maps 10–14.

———. 1856. West-Sibirien, seine Natur-Beschaffenheit Industrie und geographisch-politische Bedeutung. *Petermanns Mitteilungen* 2:201–21; maps 12, 13.

———. 1857. Siebenbürgen, physikalische-statistische Skizzen. *Petermanns Mitteilungen* 3:508–13; map plate 25.

Petermann, A., and Behm, E. 1856. Die Verbreitung der hauptsächlichsten Kultur-Produkte in den Vereinigten Staaten von Nord-Amerika. *Petermanns Mitteilungen* 2:408–39; maps 20–35.

Peterson, W. 1979. *Malthus.* Cambridge: Harvard University Press.

Peuchet. 1805. *Statistique élémentaire de la France.* Paris: Gilbert.

Playfair, W. 1786. *The commercial and political atlas; representing by means of stained copperplate charts the exports, imports and general trade of England . . . with observations . . . added charts of the revenue and debts of Ireland . . . by James Corry.* London. (2d ed., 1787; 3d ed., 1801.)

———. 1801. *The statistical breviary; shewing on a principle entirely new the resources of every state and kingdom in Europe. . . .* London.

Porter, G. R. 1837. On the connexion between crime and ignorance, as exhibited in criminal calendars. *Transactions Statistical Society of London* 1 (part 1):97–103. (Paper delivered 1835.)

Preuss, H. 1958. Johann August Zeune als Hauptvertreter der "reinen" Geographie. *Erdkunde* (Archiv für wissenschaftliche Geographie, Bonn) 12:277–84.

———. 1959. Johann August Zeune in seinem Einfluss auf Carl Ritter. *Erde* (Zeitschrift der Gesellschaft für Erdkunde zu Berlin) 90:230–40.
Prideaux, S. T. 1909. *Aquatint engraving*. London: Duckworth.
Prussia. 1828. *Administrativ-Statistischer Atlas vom Preussischen Staate*. Berlin.
Quetelet, A. 1829. *Recherches statistiques sur le Royaume des Pays-Bas*. Bruxelles: Tarlier.
———. 1831. *Recherches sur le penchant au crime aux différens âges*. Bruxelles: Hayez.
———. 1832. *Recherches sur le penchant au crime aux différans âges*. Nouveaux mémoires de l'Académie Royale des Sciences et Belle-Lettres de Bruxelles, Vol. 7.
———. 1835. *Sur l'homme et le développement de ses facultés, ou essai de physique sociale*. Paris: Bachelier.
———. 1838. *Ueber den Menschen und die Entwicklung seiner Fähigkeiten*. Trans. V. A. Riecke. Stuttgart: Schweizerbart.
———. 1842. *A treatise on man and the development of his faculties . . . now first translated into English*. Edinburgh: Chambers.
———. 1850. Letters addressed to H.R.H. the Grand Duke of Saxe-Coburg and Gotha on the theory of probabilities as applied to the moral and political sciences . . . translated from the French by O. G. Downes . . . London, 1849. *Edinburgh Review* 49:1–57.
Railway Commissioners. 1838a. *Second report of the commissioners appointed to consider and recommend a general system of railways for Ireland*. Presented to both Houses of Parliament by Command of Her Majesty. Dublin: HMSO.
Railway Commissioners. 1838b. *Atlas to accompany second report of the railway commissioners, Ireland*. N.p.
Ramsay, R. H. 1972. *No longer on the map, discovering places that never were*. New York: Viking Press.
Rawson, R. W. 1838. Abstract of the second report of Irish railway commissioners. *Journal of the Statistical Society of London* 1:257–77.
Riel, H. F. van. 1925. Pierre Ancellin. *Tijdschrift voor Kadaster en Landmeetkunde* 40:51–56, 133–44.
Ristow, W. W. 1973. *Guide to the history of cartography*. Washington, D.C.: Library of Congress.
———. 1975. Lithography and maps, 1796–1850. In Woodward (1975a), pp. 77–112.
Ritchie, G. S. 1973. Introduction. In Howse and Sanderson (1973), pp. 9–13.
Ritter, C. 1806. *Sechs Karten von Europa mit erklärendem Texte . . . .* Schnepfenthal. (2d ed., 1820; maps appeared separately as early as 1804.)
Robinson, A. H. 1955. The 1837 maps of Henry Drury Harness. *Geographical Journal* 121:440–50.
———. 1967. The thematic maps of Charles Joseph Minard. *Imago Mundi* 21:95–108.
———. 1971. The genealogy of the isopleth. *Cartographic Journal* 8:49–53. Also in *Surveying and Mapping* 32 (1972):331–38.
———. 1973. The elusive longitude. *Surveying and Mapping* 33:447–54.

———. 1975. Mapmaking and map printing: The evolution of a working relationship. In Woodward (1974a), pp. 1–23.

———. 1976a. Revolutions in cartography. *Proceedings, American Congress on Surveying and Mapping,* 36th Annual Meeting. Washington, D.C., pp. 333–39.

———. 1976b. Nathaniel Blackmore's plaine chart of Nova Scotia: Isobaths in the open sea? *Imago Mundi* 28: (2d ser., vol. 2):137–41.

Robinson, A. H., and Petchenik, B. B. 1976. *The nature of maps: Essays toward an understanding of maps and mapping.* Chicago: University of Chicago Press.

Robinson, A. H.; Sale, R. D.; and Morrison, J. L. 1978. *Elements of cartography.* 4th ed. New York: John Wiley.

Robinson, A. H., and Wallis, H. 1967. Humboldt's map of isothermal lines: A milestone in thematic cartography. *Cartographic Journal* 4:119–23.

Rothenburg, J. N. C. 1836. *Die Cholera-Epidemie des Jahres 1832 in Hamburg.* Hamburg: Perthes and Besser.

Royal Statistical Society. 1934. *Annals of the Royal Statistical Society.* London.

Schilder, G. G., ed. 1975. *Lijst van Karten in de Ackersdijck-Collectie van de Rijksuniversiteit Utrecht.* Bulletin van de Vakgroep Kartografie no. 1. Utrecht: Geografisch Instituut van de Rijksuniversiteit.

Schoen, H. H. 1938. Prince Albert and the application of statistics. *Osiris* 5:276–318.

Schouw, J. F. 1823a. *Grundzüge einer allgemeinen Pflanzengeographie.* Translated from Danish by J. F. Schouw. Berlin.

———. 1823b. *Pflanzengeographischer Atlas zur Erläutering von Schouws Grundzügen einer allgemeinen Pflanzengeographie.* Berlin: Reimer.

———. 1833a. *Europa: Physische-geographische Schilderung.* Copenhagen.

———. 1833b. *Atlas zu Schouws Europa.* Copenhagen.

———. 1839. *Tableau du climat de l'Italie.* Copenhagen: Chez Gyldendal.

Scrope, G. P. 1825. *Considerations on volcanoes ... leading to the establishment of a new theory of the earth.* London: Phillips.

———. 1827. *Memoir on the geology and extinct volcanos of central France;* with *Maps and plates to the memoir on the geology and volcanic formations of central France.* London: Longmans et al.

———. 1833. *Principles of political economy, deduced from the natural laws of social welfare, and applied to the present state of Britain.* London: Longmans.

———. 1873. *Political economy for plain people, applied to the past and present state of Britain.* 2d ed. London: Longmans, Green.

Senefelder, A. 1819. *A complete course of lithography: Containing clear and explicit instructions ... a history of lithography ....* Trans. A. S. London: Ackermann.

Seyn, E. de. 1935. *Dictionnaire biographique des science, des lettres, et des arts en Belgique.* Bruxelles.

Sheppard, T. 1920. *William Smith: His map and memoirs.* London: A.

Brown. (Reprinted from *Proceedings of the Yorkshire Geological Society*, vol. 19, pt. 3.)

Siegfried, A. 1960. *Itinéraires de contagions: Epidémies et idéologies*. Paris: Libraire Armand Colin; English translation (1965), *Germs and ideas: Routes of epidemics and ideologies*. Edinburgh and London: Oliver and Boyd.

Sinnhuber, K. A. 1959a. Alexander von Humboldt 1769–1859. *Scottish Geographical Magazine* 75:89–101.

———. 1959b. Carl Ritter 1779–1859. *Scottish Geographical Magazine* 75:152–63.

Skelton, R. A. 1958. Cartography. In *A history of technology*, ed. C. Singer et al., 4:596–628. New York and London: Oxford University Press.

———. 1965. *Decorative printed maps of the fifteenth to eighteenth centuries*. London: Spring Books.

———. 1972. *Maps: A historical survey of their study and collecting*. Chicago: University of Chicago Press.

Smith, E. C. 1972. Engineering and invention in the eighteenth century. In Ferguson (1972), pp. 92–112.

Smith, W. 1815. (Map.) A delineation of the strata of England and Wales with a part of Scotland.... London.

Snow, J. 1855. *On the mode of communication of cholera*. 2d ed. London: Churchill.

Société de Géographie de Paris. 1822. Programme des prix mis au concours dans la première assemblée générale annuelle de l'an 1822. *Bulletin* 1 (1st ser.):64–65.

Somerhausen, H. N.d. (Map.) Carte figurative de l'instruction populaire des Pays Bas. N.p. (Date probably post-1829; place probably Bruxelles.)

Stanford, E. 1904. Obituary. *Geographical Journal* 24:686–87.

Steinhauser, A. 1858. Beiträge zur Geschichte der Entstehung und Ausbildung der Niveaukarten, sowohl See- als Landkarten. *Mitteilungen der K.-K. Geographischen Gesellschaft* 2:58–74.

Stevenson, L. G. 1965. Putting disease on the map: The early use of spot maps in yellow fever. *Journal of the History of Medicine and Allied Sciences* 20:227–61.

Stommel, H. 1950. The Gulf Stream: A brief history of the ideas concerning its causes. *Scientific Monthly* 70:242–53.

Strasser, G. 1975. The toise, the yard and the metre—the struggle for a universal unit of length. *Surveying and Mapping* 35:25–46.

Sydow, E. von. 1838–40. *Physischer Wandatlas*. Gotha: Justus Perthes.

———. 1847. *E. von Sydows Schul-Atlas*. Gotha: Justus Perthes.

———. 1855. *E. von Sydows Orographischer Atlas*. Gotha: Justus Perthes.

———. 1859. Der kartographische Standpunkt Europas am Schlusse des Jahres, 1858.... *Petermanns Mitteilungen* 5:209–56.

Szaflarski, J. 1959. A map of the Tatra Mountains drawn by George Wahlenberg in 1813 as a prototype of the contour-line map. *Geografiska Annaler* 41:74–82.

Taylor, E. G. R. 1956. *The haven-finding art*. London: Hollis and Carter.
Thrower, N. J. W. 1969a. Edmond Halley and thematic geo-cartography. In *The terraqueous globe*. Los Angeles: William Andrews Clark Memorial Library, University of California.
———. 1969b. Edmond Halley as a thematic geo-cartographer. *Annals, Association of American Geographers* 59:652–76.
———. 1972. *Maps and man*. Englewood Cliffs, N.J.: Prentice-Hall.
———, ed. 1978. *The compleat plattmaker: Essays on chart, map, and globe making in England in the seventeenth and eighteenth centuries*. Los Angeles: University of California Press.
Tobler, W. 1973. Choropleth maps without class intervals? *Geographical Analysis* 3:262–65.
Tooley, R. V. 1952. *Maps and map-makers*. 2d ed. London.
Torrey, H. B. 1938. Athanasius Kircher and the progress of medicine. *Osiris* 5:246–75.
Troll, C. 1960. The work of Alexander von Humboldt and Carl Ritter: A centenary address. *Advancement of Science*, no. 64, pp. 441–52.
Ungar, E. 1935. Ancient Babylonian maps and plans. *Antiquity* 9:311–22.
———. 1937. From cosmos picture to world maps. *Imago Mundi* 2:1–7.
Verner, C. 1974. Carto-bibliographical description: The analysis of variants in maps printed from copperplates. *American Cartographer* 1:77–87.
———. 1975. Copperplate printing. In Woodward (1975a), pp. 51–75.
Walckenaer, C.-A. 1830. *Vies de plusieurs personnages célèbre des temps ancien et moderne*. Laon.
Wallis, H. 1973. Maps as a medium of scientific communication. *Etudes d'histoire de la Géographie et de la Cartographie*. Wrocław, Warsaw, Krakow, Gdansk: Zaclad Narodowy Imienia, Ossolinskich Wydawniatwo Poliskiej Akademii Nauk, pp. 251–62.
Ward, D. 1971. Living in Victorian towns. *Geographical Magazine*, May, pp. 574–81.
Weller, E. 1911. August Petermann: Ein Beitrag zur Geschichte der geographischen Entdeckungen und der Kartographie im 19. Jahrhundert. *Quellen und Forschungen zur Erd- und Kulturkunde* 4:1–284.
Wilde, W. R. 1843. Report upon the tables of deaths, part V, Special sanitary report on Dublin City. Miscellaneous tables, Part 4 of census of Ireland 1841, pp. 68–74. (Map ff.)
Wolkenhauer, W. 1916–17. Aus der Geschichte der Kartographie. *Deutsche Geographische Blätter* (Bremen) 38:101–28, 157–201.
Wolter, J. A. 1972. The heights of mountains and the lengths of rivers. *Quarterly Journal of the Library of Congress* 29 (no. 3):186–205.
Woodbridge, W. C. 1826. *School atlas to accompany Woodbridge's rudiments of geography / Atlas on a new plan....* Hartford, Conn.: O. D. Cooke.
Woodward, D., ed. 1975a. *Five centuries of map printing*. Chicago: University of Chicago Press.
———. 1975b. The woodcut technique. In Woodward (1975a), pp. 25–50.

———. 1978. English cartography, 1650–1750: A summary. In Thrower (1978), pp. 159–93.

Wright, J. K. 1925. *The geographical lore of the time of the crusades.* Research Series no. 15. New York: American Geographical Society. New York (reprinted New York: Dover Publications, 1965).

———. 1955. "Crossbreeding" geographical quantities. *Geographical Review* 45:52–65.

———. 1966. *Human nature in geography.* Cambridge: Harvard University Press.

Wyld, J. 1815. (Map.) Chart of the world shewing the religion, population and civilization of each country. London.

Zeune, A. 1808. *Gea. Versuch einer wissenschaftlichen Erdbeschreibung.* Berlin. (2d ed., 1811.)

———. 1815. *Erdansichten oder Abriss einer Geschichte der Erdkunde vorzüglich der neuesten Fortschritte in dieser Wissenschaft.* Berlin. (2d ed., 1820.)

———. 1830. *Gea. Versuch, die Erdrinde sowohl in Land-als Seeboden mit Bezug auf Natur und Völkerleben zu schildern.* 3d ed. (with four maps). Berlin.

Zimmermann, E. A. W. 1777. *Specimen zoologiae geographicae, quadrupedum domicilia et migrationes sistens.* Leiden: Theodorum Haak et Socios.

———. 1778–83. *Geographische Geschichte des Menschen und der allgemein verbreiteten vierfüssigen Thiere.* 3 vols. Leipzig: Weyganschen Buchhandlung. ("Zoologischen Weltcharte" in vol. 3, 1783).

Zögner, L. 1979. *Carl Ritter in seiner Zeit. Austellungskataloge 11.* Berlin: Staatsbibliothek Preussischer Kulturbesitz.

# INDEX

References to reproductions of maps are printed in boldface type.

Académie des Sciences (Paris), 18; history and publications of, 35–36
*Administrativ-Statistischer Atlas vom Preussischen Staate*, 65, 113, **114,** 199
Air pressure. *See* Atmospheric pressure, mapping of
Ancellin, Pierre (1653–1720), 210
Angeville, A., comte d' (1796–1856), 115–17, **116,** 162, **163,** 182, **183**
Animals, mapping distributions of. *See* Zoology, mapping of
Aquatint, 23, **76, 117,** 195, **196, 197**
Area symbols, 189
Arrowsmith, A. (1750–1833), 81
*Asia Polyglotta nebst Sprachatlas,* 54, 132, **133**
Atlas: by Berghaus, H. (see *Physikalischer Atlas* . . .); by Blaeu, J., 137; by Block, M. (see *Puissance comparé des divers états de l'Europe—Atlas*); by Guettard, J. E., and Monnet (see *Atlas et description minéralogiques de la France*); by Johnston, A. K. (see *National Atlas* . . . , *The* [1843], *Physical Atlus* . . . *The* [1848], *Atlas of Physical Geography, Illustrating* . . . [1852]); by Klaproth, J. (see *Asia Polyglotta nebst Sprachatlas*); by National Society (see *Maps Illustrative of the Physical, Political and Historical Geography* . . . ; by Olsen, O. N. (see *Atlas pour le tableau du climat d'Italie*); by Petermann, A. (see *Atlas of Physical Geography with Descriptive Letterpress*); of Prussia (see *Administrativ-Statistischer Atlas vom Preussischen Staate*); by Railway Commissioners, Ireland (see *Atlas to Accompany Second Report of the Railway Commissioners* . . .); by Ritter, C. (see *Sechs Karten von Europa*); by Schouw, J. F. (see *Pflanzengeographischer Atlas* . . .); by Stieler, A., 65–66, 124; thematic economic, development of, 146–47; thematic physical, development of, 64–67
*Atlas et description minéralogiques de la France,* 64, 91, **92**
*Atlas of Physical Geography with Descriptive Letterpress* . . . , 75–78, **79, 83**
*Atlas of Physical Geography, Illustrating* . . . , 100, **101,** 180
*Atlas pour le tableau du climat d'Italie,* **74,** 75
*Atlas to Accompany Second Report of the Railway Commissioners* . . . , 117–20, **117,** 147, **149,** 204–5, 208–9, **209**
Atmospheric pressure, mapping of, 73–75

Baker, Robert, 172, 186, **187**
Balbi, A. (1782–1848), 62, 158, **159,** 199
Base maps for thematic mapping, 17–22

Beaumont, Elie de (1798–1874), 91
Belpaire, Alphonse (1817–54), 147, 149–50, 208
Berghaus, Heinrich (1797–1884): *Geographische Kunstschule*, 65; maps of characteristics of people by, 134–37; origin of *Physikalischer Atlas* by, 65–67; population density map by, 114, **115**; relation of, to A. K. Johnston, 66; relation of, to O. N. Olsen, 98, **99**; use of layer tints by, **77**, 214; use of shading by, 200. See also *Physikalischer Atlas*
Berlin Industrial Exhibition (1844), maps for, 141
Beste, George (?–1584), 11
Biology, mapping of. See Botany, mapping of; Zoology, mapping of
Blackmore, Nathaniel (?–1716), 210
Block, Maurice (1816–1901), 65, 130, **131**, 137, 146–47, **148**
Blum, H., 141
Bone, W., 125, 205
Booth, Charles (1840–1916), 186, **188**
Borri, Christoforo (fl. ca. 1620), 84
Botany, mapping of, 60–62, **61**, 101–3, **103, 104, 105**
Boue, Ami (1794–1881), 91
Brandes, Heinrich Wilhelm (1777–1834), 73–74
Bredsdorff, J. H. (1790–1844), 97, **98, 99**, 213, **215**
British Association for the Advancement of Science, 36, 182
Bromme, Traugott (1802–66), 225 n.114
Brongniart, Alexandre (1770–1847), 58–59, **59**, 91
Bruinsz, Pieter (fl. ca. 1580), 210, **211**
Buache, Philippe (1700–1773), 52–54, 59; career of, 87; use of isolines by, 87–90, **88, 89, 90,** 210
Buch, Leopold F. von (1774–1853), 91
Büsching, A. F. (1724–93), 31

Calculus, differential, 28
Carte de Cassini, 19, 21, 64
*Carte figurative*, 151–53, 231 n.273
Carte géométrique de la France, 19–21
Cartography: church-oriented symbology in, 10–11; early knowledge of, 2; qualitative compared with quantitative representation in, 189, 194; revolutions in, 12–15; as substitute for space, 1; symbolism in, 189–90; thematic, compared with general, 16–17, 69; thematic, development of, 17; topographic, development of, 19–20; treatment of transitions and intermixtures in, 139; use of color in, 102, 139, 142, 172; use of shading in, 110
Cary, William (1759–1825), 58
Cassini, Giovanni Domenico (Jean Dominique) (1625–1712), 19
Cassini, Jacques (1677–1756), 19, 31
Cassini, Jacques-Dominique, comte de (1748–1845), 19, 93–94
Cassini de Thury, César François (1714–84), 19
Census: of Great Britain (1851), 125–28, **127, 198**, 205; of Ireland (1841), 119–20, **120**, 162–63, 182–86, **184, 185**
Censuses, development of, 34
Circles, proportional. See Proportional point symbols, circles as
Compass variation, mapping of, 51, 83–86. See also Halley, Edmond; Kircher, Athanasius
Chadwick, Edwin (1800–1890), 186
Charpentier, J. F. W. (1728–1805), 91
Charts, 7
Chemistry, developments in, 28
Cholera epidemics, mapping associated with, 64, 156, 170–72, **171**, 174–78, **177, 179**, 180–81, **181**
Choropleth map, 111, 113, 115–16, 160, 198–200; class limits in, 199
Clark, Samuel (1810–75), 78, **80**, 124–25, **126**, 142–43, **143**; career of, 125
Classes in mapping statistical data, 115, 162, 164, 199
Color: logical use of, 139; printing of, 23–24, 108, 193
Communication among cartographers, 190
Compass, invention of, 84
Compass variation, mapping of, 46, 83–86, **85**
Continuous tone, 196–98. See also Shading, variable
Contour, 202; history of, 210–19; shaded, 174
Copperplate engraving: aquatint in,

195; color printing in, 193; compared with lithography, 192–93; as a duplication technique, 23, 191–98; production of tones and shading in, 23, 192, 196
Crayon: shading by, 23, 195; use in lithography, 195, 198
Crime, mapping of, 158–69, **159, 161, 163, 165, 167, 168, 169**
Crome, August Friedrich Wilhelm (1753–1833), 33, 54–55, **56,** 141, 206
Cruquius, Nicolaas Samuelsz (1678–1754), 210
Curve line, use of by E. Halley, 84–86. *See also* Isoline
Cuvier, Georges (1769–1832), 29, 58–59, **59,** 91

Dampier, William (1652–1715), 69
Dasymetric map, 113, 118–19, 199
Dead reckoning, 21
De Brahm, J. G. W. (1717–99), 81
Declination. *See* Compass variation, mapping of
Delaméthrie, 81
Delisle, Guillaume (1675–1726), 87
Density: concept of, 216; portrayal of, 198–202
D'Halloy. *See* Omalius d'Halloy, Jean Baptiste Julien d'
Differential calculus, development of, 28
Dot map, 110–13, 160, 170, 175, 200–202
Dower, John, 76, 196
Drummond, Thomas (1797–1840), 118
Du Carla, Marcellin (1738–1816), 94, 211, **212**
Dufour, Guillaume Henri (1787–1875), 94
Dufrenoy, P. A. (1792–1857), 91
Dupain-Triel, J. L. (1722–1805), 94–95, **95,** 210, 213, **214**
Duperrey, L.-I. (1786–1865), 86
Dupin, Charles (Baron) (1784–1873), 62, 116–17, 144, 156–58, **157,** 199
Duplication techniques: 22–24. *See also* Copperplate engraving; Lithography; Printing; Woodcut; Wood engraving

Earth, shape of, 19, 28
Eclipse of the sun, map of, 49–50, **50**

Ecole des Ponts et Chaussées, 37
Ecole Polytechnique, 37
Economic activities, development of maps of, 140–44
Economic data, mapping of, 54–55, **56,** 140–47, **143, 145, 146, 207**
Educational institutions, relation of to scientific developments, 37
Encyclopedias, 38
Environment, growth of interest in, 39–40
Etching. *See* Aquatint
Ethnology, mapping of, 137–40, **139**
Etzlaub, Erhard (ca. 1460–1532), 7
Exhibition, Berlin Industrial (1844), maps for, 141
Expense: of copperplate engraving, 191; of lithography, 192–93

Flat tone, 196
Fletcher, Joseph (1813–52), 121, **122,** 164–66, **165**
Flow map, 64, 147–54, 208–9
Forsell, Carl Gustav (1783–1848), 213
Forster, Georg (1754–94), 155
France, topographic mapping of, 18–21
Franklin, Benjamin (1706–90), 81, **82**
Füchsel, Georg C. (1722–73), 54, 90–91

Geognosy. *See* Geology
*Geographische Kunstschule,* 65, 124
Geological Society of London, 86
Geology: developments in, 28–30; mapping of, 52–60, **53, 59,** 86–92, **88, 89, 90, 92, 93;** relation to canals and mining, 57; surveys of, 86
Geomagnetism, 46, 51, 57, 83–86, **85**
Gläser, F. Gottlob (1749–1804), 54, 91
Graduated circles. *See* Proportional point symbols
Greenough, G. B. (1778–1855), 91
Greg, W. R., 161–62
Griffith, Richard (1784–1878), 118
Grounds, etching, in aquatint, 195
Guerry, André-Michel (1802–66), 62–63, 144, 156, 158, **159,** 160–61, 166, **168, 169,** 199
Guettard, Jean-Etienne (1715–86), 52–54, 59, 64, 87, **92**
Gulf Stream, 81, **82**

Hachures, 208, 213
Halley, Edmond (1656–1742): career

of, 46–47; charts of compass variation by, 84–86, **85;** chart of winds by, **12,** 69–70, **70–71;** map of sun's eclipse by, **50;** map of tides by, 49, 82; use of curve line by, 51
Hansteen, Christopher (1784–1873), 86
Happel (Happelius), Eberhard Werner (1647–90), 46, **47**
Harness, Henry Drury (1804–83): flow maps by, 147, **149,** 208, **209;** population map by, 118–19, **117, 196,** 199; use of proportional symbols by, 204–5
Harrison, John (1693–1776), 21
Hauslab, Ritter von, 213
Hensel, Gottfried (fl. 18th C.), 54, 55, 130, **132,** 137
Hereford Map, 11–12
Housing characteristics: mapping of, 182–84, **183, 184;** taxation of, 182
Hubertz, J. R., 175
Humboldt, Alexander von (1769–1859): career of, 67, 155; introduction of isotherm by, 60, 70–72, **72;** *Kosmos* by, 65, 67; measurement of mountain heights by, 94; as "morphologist of nature," 29–30; role of, in *Physikalischer Atlas,* 65; scientific renown of, 60
Humboldtian sciences, 30
Hume, Abraham, 134
Hutton, James (1726–97), 29
Huygens, Christiaan (1629–95), 18–19, 31
Hyetographic map. *See* Precipitation, mapping of

Industrial Exhibition, Berlin (1844), maps for, 141
Industrial revolution, 40–41
International Exposition, London (1851), 79; map for, 142
Inventions, growth in numbers of, 41
Isobar, 74–75
Isobarometric line, 74
Isobath, 87, 202, 210
Isochimenes, 72
Isogone, 226 n.156
Isoline: in biological mapping, 216; concept of, 210; as curve line, 51, 84–86, 226 n.154; Humboldt's introduction of isotherm, 60; isobarometric line, 74; isobath, 87, 202, 210; isochimenes, 72; isogone, 226 n.156; isopleth, 111, 128–30, 215–18; isotheres, 72; isotherm, 60, 71–72; proliferation of terms for, 72, 216; use of by Buache, 87–90
Isopleth, 111, 128–30, 215–18; naming of, 217–18
Isotheres, 72
Isotherm: acceptance of, 71–72; introduction of by Humboldt, 60

Johnston, Alexander Keith (1804–71): interest of, in Berghaus's *Physikalischer Atlas,* 66; maps of landforms by, 100, **101;** preparation of *The Physical Atlas* by, 66; rain map by, 75, **197;** use of Olsen's contour map of Europe by, 98, **99,** 211; zoological maps by, 105–6, **106, 107.** See also *National Atlas . . . , The; Physical Atlas . . . , The; Atlas of Physical Geography, Illustrating . . .*

Kämtz, Ludwig (1801–67), 74, 216
Kircher, Athanasius (1602–80): career of, 45; map of compass variation by, 46, 84; maps of ocean currents by, 45, 80–81, **81**
Klaproth, Julius (1783–1835), 54, 132, **133**
Kombst, Gustav (?–1846), 137–40, **138**
*Kosmos,* 65, 67; atlas for, 225 n.114
Kutscheit, Johann Valerius (fl. ca. 1840), 134–35, **134, 135,** 141

Lalanne, Léon (1811–92), 216–17
Land surface, mapping the, 92–101, **95, 96, 100, 101,** 208, 211–13
Lange, Heinrich (1821–93), 66, 124
Languages, mapping of, 54, **55,** 130–32, **132, 133**
Larcom, Thomas A. (1801–79), 18, 182, 213; career of, 120
Layer tint map, 95, **95,** 130, 210–14
Legend, for proportional circles, 205
Lehmann, Johann Georg (1765–1811), 208
*Le Neptune françois,* 19
Length, quest for universal standard of, 30–32
Le Roy, Pierre (1717–85), 22
Libraries, European, map of numbers of books in, 187

Line symbols, 189
Lister, Martin (1638?–1712), 51–52
Lithography: color printing in, 193; comparative expense of, 110, 192; compared with copperplate engraving, 110, 192–93; discovery of, 57, 224 n.93; as duplication technique, 23, 192–98; engraving in, 192; production of tones and shading in, 23, 192, 198; use of crayon in, 195, 198
Living conditions, mapping of, 182–88, **183, 184, 185, 187, 188**
London International Exposition (1851), 79; map for, 142
Longitude: determination of, 20–22; search for, 51
Loomis, Elias (1811–99), 75, 78
Lyell, Charles (1797–1875), 91

Macrocosm-microcosm analogy, 52
Magnetic declination. *See* Compass variation, mapping of
Mahlmann, Wilhelm (1812–48), 71
Malgaine, Joseph-François (1806–65), 174, **175**
Mappae mundi, 11
Mapping: of atmospheric pressure, 73–75; of botany, 60–62, **61**, 101–3, **103, 104, 105;** of compass variation, 46, 83–86, **85;** of crime, 158–69, **159, 161, 163, 165, 167, 168, 169;** of eclipse of the sun, 49–50, **50;** of economic data, 54–55, **56,** 140–47, **143, 145, 146, 207;** of ethnology, 137–40, **139;** of geology, 52–60, **53, 59,** 86–92, **88, 89, 90, 92, 93;** of geomagnetism, 46, 51, 57, 83–86, **85;** of the land surface, 92–101, **95, 96, 100, 101;** of languages, 54, **55,** 130–32, **132, 133;** of living conditions, 182–88, **183, 184, 185, 187, 188;** of medical data, 40, 64, 170–81, **173, 175, 176, 177, 179, 180, 181;** method of, for statistical data, 110–11; of mineral waters, 174; of moral statistics, 35, 41, 62–64, **63,** 156–70, **157, 159, 161, 163, 165, 167, 168, 169;** of occupations, 142, **143;** of ocean currents, 45–46, **47,** 80–81, **81,** 82, 83; of passenger traffic, 147, **149,** 152, 153, **209;** of population, 111–30, **114, 115, 116, 117, 120, 122,** 128–30, 131, **198, 202, 204;** of population characteristics, 54, **55,** 130–40, **132, 133, 134, 136, 138;** of precipitation, 74–80, **74, 76, 77, 79, 80;** of religion, 133–37, **134, 136, 204;** of temperature, 70–72, **72, 73,** 83, **83;** of transportation, 64, 147–54, **149, 151, 152, 153;** of volcanoes, 91–92, **93;** of weather, 78–79; of winds, 47–49, **48,** 69–70, **70–71,** 75; of zoology, 62, 101, 104–6, **106, 107.** *See also* Thematic mapping
Maps: as aid to travel, 6–7; base, development of, 17–22; coloring of, by hand, 24; color printing of, 23–24, 108, 193; for figurative expression, 7–11; functions of, 3–17; general or reference, 3–6, 18–20; printing of, techniques for, 22–23, 190–98; thematic, character of, 15–17, 69
*Maps Illustrative of the Physical, Political and Historical Geography . . . ,* 65, 78–79, **80,** 125, **126,** 142–43, **143**
Map types: choropleth, 111, 113, 115–16, 160, 198–200; contour, 210–19; dasymetric, 113, 118–19, 199; dot, 110–13, 160, 170, 175, 200–202; flow, 147–54, 208–9; layer tint, 210–14; isopleth, 111, 128–30, 215–18; shaded, 75–79, 110, 121, 123, 127–28, 160, 171, 180, 200
Martini, Martinus (1614–61), 84
Mathematics, developments in, 28
Mayhew, Henry (1812–87), 166, **167**
Mechanical ruling, 23
Medical data, mapping of, 40, 64, 156, 170–81, **173, 175, 176, 177, 179, 180, 181**
Medical geography, interest in, 40
Mentzer, Thure Alexander von (1807–92), 201
Mercator's projection, 69
Metric system, development of, 31–32
Mezzotint, 23, 196, **197,** 236 n.360
Milner, Thomas (?–1882), 107
Montizon, A. Frère de, 64, 112–13, **112**
Moral statistics, mapping of, 35, 41, 62–64, **63,** 156–70, 157, **159, 161, 163, 165, 167, 168, 169**
Michaelis, Ernst Heinrich (1794–1873), 174–75, **176**
Minard, Charles Joseph (1781–1870): career of, 144; map of population by, 199; maps of movement by,

147–54, **151, 152, 153;** maps of market data by, **145, 146;** use of sectored proportional circles by, 145, **145,** 207–9, **207**
Mineralogy, development of, 29
Mineral waters, mapping of, 174
Mining, growth of, 41
Monnet, Antoine-Grimoald (1734–1817), 64

*National Atlas...*, *The*, 137–40, **138**
Natural history, developments in, 29–30
Nosogeography. *See* Medical data, mapping of

Occupations, mapping of, 142, **143**
Ocean currents, mapping of, 45–46, **47,** 80–83, **81, 82, 83**
Olsen, Oluf Nikolay (1794–1848): career of, 97; collaboration of, with J. H. Bredsdorff, 97; contour map of Europe by, 97–99, **98;** hyetographic map by, **74,** 75; entry of, in mountains of Europe competition, 97; prototype layer tint map by, 213, **215**
Omalius d'Halloy, Jean Baptiste Julien d' (1783–1875), 91

Packe, Christopher (1686–1749), 52, **53,** 86
Passenger traffic, mapping of, 147, **149,** 152, **153, 209**
Perry, Robert, 173–74, **173**
Petermann, August (1822–78): *Atlas of Physical Geography...* by, 75–78, **79, 83;** career of, 124; cholera map by, 64, 180–81, **181;** first map by, 141–42; *Geographische Mitteilungen* by, 124; map of occupations by, 142, **143;** population maps by, 114, 120, 123, **123, 127, 198,** 200, **202,** 205; stay of, in Edinburgh, 66; use of pseudo-isolines by, 217
*Peutingerische Tafel*, 6
*Pflanzengeographischer Atlas...*, 65, 102–3, **103**
Photography, effect of on thematic mapping, 222 n.35
*Physical Atlas...*, *The*, 75, 98–99, **99,** 105–7, **105, 106, 107, 134,** 138, **195, 197;** relation of, to Berghaus's *Physikalischer Atlas*, 65–67. *See also* Johnston, Alexander Keith
*Physikalischer Atlas...*, 65–67, 70, **73, 75, 76, 77,** 91, 94, 98, **99,** 103, **104,** 105, 114, 115, 134–37, **136,** 138, 180, 201, 214; connection of, with A. K. Johnston, 66; contents of, 66; origin of, 65. *See also* Berghaus, Heinrich
Picard, Jean (1620–82), 31
Pie chart, 206–8
Plants, mapping distributions of. *See* Botany, mapping of
*Plastik*, 94
Playfair, William (1759–1823), 33, 161, 206
Point data, portrayal of, 202–9
Point symbols, 189
Political arithmetic, 33
Periodicals, scholarly, 37
Poor Law Commissioners, 186
Population, mapping of, 111–30, **114, 115, 116, 117, 120, 122,** 128–30, 131, **198, 202, 204**
Population characteristics, mapping of, 54, **55,** 130–40, **132, 133, 134, 136, 138**
Portolan charts, 7, 14
Precipitation, mapping of, 74–80, **74, 76, 77, 79, 80**
Pressure. *See* Atmospheric pressure, mapping of
Printing: of color, 23–24, 108, 193; development of, 22–24; methods of, for maps, 22–23, 190–98; proliferation of methods of, in 19th century, 193; registry in, 24
Pritchard, J. C. (1786–1848), 137
Probability, study of, 28, 34–35
Profiles, of mountains, 70, 102
Proportional point symbols: circles as, 111, 119, 126–28, 206–8; judgment of, 208; legend for, 205; sectored, 206–8; use of, 203–8
Ptolemy (Ptolemaeus), Claudius (ca. 87–150), maps attributed to, 2, 4–5, 14–15
*Puissance comparé des divers états de l'Europe—Atlas*, 65, 130, **131,** 137, 146–47, **148**
Pym, William (1772–1861), 172

Quantitative linear data, portrayal of, 208–9

Quetelet, L. Adolphe J. (1796–1874): application of probability to social affairs by, 35; maps of moral statistics by, 63, 160–62, **161;** proposal of, for mapping simultaneous blooming dates, 216; translation of works of, 144; use of shading by, 200

Railways: development of, 42; growth in mileage of, 42; map of mileage of, **148**
Rank ordering data, 115, 162, 166
Rau, Carl Ferdinand von (1783–1833), 113
Ravn, Nils Frederik (1826–1910), 128–30, **129,** 217–18
Registry in printing, 24, 193
Relative population. *See* Density; Population, mapping of
Religion, mapping of, 133–37, **134, 136, 204**
Ritter, Carl (1779–1859): biological maps by, 61–62, 101–2, 105; first thematic atlas by, 64; maps of ethnology by, 137; map of land surface of Europe by, 70, **96,** 197; map of population by, 111, 203
Rockers for tone production, 194
Rollers for tone production, 194
Romanticism, growth of, 39
Rothenburg, J. N. C., 171, **171**
Roulettes for tone production, 194
Royal Society of London, establishment of, 36
Ruling machine, 192, 194

Sabine, Edward (1788–1883), 86
Sanitary map, 40
Santa Cruz, Alonso de (ca. 1500–1572), 84
Schnurrer, Friedrich (1784–1833), 179
Schouw, Joakim Frederik (1789–1852), 61, 65, 102–3, **103,** 105
Schwartzer, Ernst von, 141
Science, transformation of, from 17th to 19th century, 27–30
Scrope, George J. Poulett (1797–1876), 64, 91–92, **93,** 113–15, 199
*Sechs Karten von Europa* . . . , 61–62, 64, 70, **96,** 101–2, 111, 137, **197,** 203
Sectored circles, 206–8
Semenov-Tian-Shansky, Benjamin, 199

Senefelder, Aloys (1771–1834), 23, 192, 224 n.93
Seutter, G. Matthäus (1678–1757), 140–41
Shading, variable: production of, 23, 127–28, 160–61, 173–74, 180–81, 192–98; as technique, 110
Smelting, growth of, 41
Smith, William (1769–1839), 29, 58–60, **60**
Snow, John (1813–58), 175–79, **177, 179**
Social affairs: growth of interest in, 32–35; inhibiting factors in study of, 33–34
Société de Géographie de Paris, 96–97
Societies: scholarly, 35–38; scientific, 35–37; statistical, 36
Somerhausen, H., 62, **63,** 158
Soundings, 202
Specific population. *See* Density; Population, mapping of
Spot map. *See* Dot map
Stanford, Edward (1827–1904), 124–25
Statistical data: arrangement of, 34; classes in, 115, 162, 164, 199; conversion of, to density, 109–10, 198; graphic representation of, 33, 205–8; making of, comparable, 109–10, 158; methods of mapping of, 110–11; rank ordering of, 115, 162, 166; social, compared with physical, 62
Statistical Society of London, 222 n.54
Statistical surface, 128, 201–2, 210, 218
Statistics, development of as field of study, 28, 32–35, 155
Stieler, Adolf (1775–1836), 65
Sydow, Emil von (1812–73), 100, **100,** 130, 213, 217–18
Symbolism: need for new, 189–90; use of color in, 193

Taxation of housing characteristics, 182
Teixeira, Luis (1564–1604), 84
Temperature, mapping of, 70–72, **72, 73,** 83, **83**
Thematic mapping: base maps for, 17–22; beginnings of, 17; changes in rate of growth of, 44; development of, 26; effect of lithography on, 193; effect of photography on, 222 n.35; "golden age" of, 24; nature of, 15–17; need of new symbolism for,

189; rate of development of classes of, 147; representational innovations in, 190, 198–219; sign of maturity in, 142–43; of statistical data, 110–11; tones and color in, 23–24, 102, 108, 110; as tool for hypothesis formulation and testing, 156, 166, 170–79; use of woodcut for, 191

Thomson, George J. P. *See* Scrope, George J. Poulett

Tints. *See* Tone

Toise, length of, 31–32

T–O map, 10

Tone: continuous, 196–97, 200; flat, 196; production of, in printing processes, 23, 192–98; use of, to show quantity, 76. *See also* Shading, variable

Traffic. *See* Passenger traffic, mapping of

Transportation: advances in, 41–43; mapping of, 64, 147–54, **149, 151, 152, 153**

Variation, compass. *See* Compass variation, mapping of

Volcanoes, mapping of, 91–92, **93**

Wahlenberg, Göran (1780–1851), 61, 213

Weather, mapping of, 78–79

Weiland, Carl Ferdinand (1782–1847), 137, 174

Werner, Abraham Gottlob (1750–1817), 29

Whiston, William (1667–1762), 51, 86

Wilcke, Johan Carl (1732–96), 51, 86

Wilde, William Robert Wills (1815–76), 184–86, **185**

Winds, mapping of, 12, 47–49, **48,** 69–70, **70–71,** 75

Woodcut, 22; comparison of, with wood engraving, 235 n.345; printing of, with text, 191

Wood engraving, 22, 191; comparison of, with woodcut, 235 n.345

Wyld, James (the elder) (1790–1836), 111, 133, 203–4, **204**

Yellow fever, maps of, 170

Zeune, August (1778–1853), 238 n.400

Zimmermann, Eberhard August Wilhelm (1743–1815), 62, 104–5

Zoology, mapping of, 62, 101, 104–6, **106, 107**

65946

| GA | Robinson, Arthur |
| 201 | Howard, 1915- |
| .R63 | Early thematic mappin[g] |
| 1982 | in the history of |
| | cartography |